普通高等教育工程理念与创新应用系列教材

中国电力教育协会高校电气类专业精品教材

电气工程及其自动化专业英语教程

第二版

凌跃胜　宋桂英　李永建　　编

郝晓弘　主审

中国电力出版社
CHINA ELECTRIC POWER PRESS

内 容 提 要

本书是针对高等工科院校电气工程及其自动化专业科技英语阅读课程的需要而编写的，内容涵盖电气工程中电机与电器、电工理论与新技术、电力电子与电力传动、电力系统及其自动化、高电压与绝缘技术等5个二级学科相关专业基础知识。同时，考虑到电气工程学科新兴技术领域的发展，从不同角度选用了相关专业著作与文献资料。全书按二级学科分为5个部分，共计30个单元，每单元包含课文、专业英语词汇、注释和综合练习等内容。为了提高使用者的科技英语阅读理解能力、翻译能力和写作能力，在教程的附录部分增加了专业英语阅读、翻译和写作常识，可根据教学需要选用。

本书主要作为高等学校电气工程及其自动化专业教材，也可作为科技人员工程技术人员学习专业英语的参考用书。

图书在版编目（CIP）数据

电气工程及其自动化专业英语教程 / 凌跃胜，宋桂英，李永建编．—2版．—北京：中国电力出版社，2021.8（2022.9重印）

ISBN 978-7-5198-5741-7

Ⅰ.①电… Ⅱ.①凌… ②宋… ③李… Ⅲ.①电工技术－英语－高等学校－教材②自动化技术－英语－高等学校－教材 Ⅳ.①TM②TP2

中国版本图书馆CIP数据核字（2021）第129552号

出版发行：中国电力出版社
地　　址：北京市东城区北京站西街19号（邮政编码100005）
网　　址：http://www.cepp.sgcc.com.cn
责任编辑：牛梦洁
责任校对：王小鹏
装帧设计：赵丽媛
责任印制：吴　迪

印　刷：望都天宇星书刊印刷有限公司
版　次：2007年8月第一版　2021年8月第二版
印　次：2022年9月北京第十次印刷
开　本：787毫米×1092毫米　16开本
印　张：16
字　数：387千字
定　价：45.00元

前　言

专业英语是高校学生专业培养不可缺少的一部分，其重要性不言而喻。为了适应电气工程及其自动化专业人才培养的需要，作者根据多年从事高等工科院校电气工程及其自动化专业英语课程的教学实践，结合教育部的专业目录要求编写了《电气工程及其自动化专业英语教程》，于2007年出版。

自教材出版后，得到的普遍反映是：教材内容覆盖面广、形式新颖，通过学习，能够掌握电气工程领域的基本专业词汇，能够顺利地阅读本专业英文资料，同时还能了解电气工程领域的相关知识和方向，效果良好。根据本书十多年的教学使用情况，考虑到电力新技术、新理论的迅速发展，结合教育部最新颁布的专业培养要求，作者决定对该教材进行修订。再版教材保持专业覆盖面较广的特色，结构体系不变，在内容上删除相对陈旧、使用较少、涉及面窄的内容，增添近年来在电气工程领域中应用广泛、发展迅速的前沿知识以及对学生有吸引力的英文原文资料，以适应电气工程及其自动化专业工程认证和新工科对本科教学的要求。

本书内容涵盖电气工程及其自动化专业的电机与电器、电力电子与电力传动、电力系统及其自动化、电工理论与新技术、高电压与绝缘技术各个领域，全书按二级学科分为5个部分，共计30个单元，每单元包含课文、专业英语词汇、注释和综合练习等内容。

本书由从事多年专业英语教学的凌跃胜、宋桂英、李永建教授编写，郝晓弘主审。教材修订过程中李珊瑚博士、汤雨教授、辛振教授、黄珺博士、胡艳芳博士、王笑雪博士、张长庚博士等提供了大量原始资料，并参与了教材编写修订工作，在此表示衷心感谢！

编者

2021年3月

第一版前言

为贯彻落实教育部《关于进一步加强高等学校本科教学工作的若干意见》和《教育部关于以就业为导向深化高等职业教育改革的若干意见》的精神，加强教材建设，确保教材质量，中国电力教育协会组织制订了普通高等教育“十一五”教材规划。该规划强调适应不同层次、不同类型院校，满足学科发展和人才培养的需求，坚持专业基础课教材与教学急需的专业教材并重、新编与修订相结合。本书为新编教材。

科学技术的迅速发展与国际学术交流的增多，对高等院校学生专业英语的阅读、翻译和写作能力都提出了更高的要求，各院校所设置的专业英语课程作为专业英语基本技能训练，也越来越显示出其重要性。为了贴近新技术、新理论的不断发展和应用，目前使用的电气工程及其自动化专业英语教材需要更新和增加内容，以满足电气工程及其自动化专业的人才培养需要。作者根据多年从事高等院校电气工程及其自动化专业英语课程的教学实践，结合国家教委最新颁布的专业目录要求，编写了电气工程及其自动化专业的专业英语教材。

为了在选材上力求先进性，本书大部分内容材料选自国外相关专业的参考书，在编排上系统地贯穿了电气工程及其自动化专业的专业基础课程，并添加和更新了部分英语教材内容，补充了许多与工程实践相关的新知识，如电磁兼容技术、现代电力电子技术以及高压直流传输等，以达到在学习专业外语的同时，丰富和补充电气工程及其自动化专业新理论、新知识的目的。

本书由凌跃胜教授、宋桂英副教授、黄文美副教授共同编写。书中第一、二部分由黄文美编写，第三、四部分由宋桂英编写，第五部分以及附录由赵争菡编写，全书由凌跃胜教授统稿。兰州理工大学的郝晓弘教授审阅了全书，并提出了大量的宝贵意见和建议。河北工业大学杨晓光副教授、安金龙副教授，在国外工作的何仁杰博士、叶秋波博士等，为本书编写提供了很多宝贵的资料，在此表示衷心感谢。

由于编者水平有限，书中难免存在不足之处，殷切期望广大读者批评指正。

编者

2007 年 1 月

目　录

[illegible]

[illegible]

Part Ⅰ Electrical Machines and Electrical Apparatus

Unit 1 Power Transformer

Text A Construction and Principles of Power Transformer

Transformer is an indispensable component in electric energy conversion systems. It makes possible electric generation at the most economical generator voltage, power transfer at the most economical transmission voltage, and power utilization at the most suitable voltage for the particular utilization device. The transformer is also widely used in low-power, low-current electronic and control circuits for performing such functions as matching the impedances of a source and its load for maximum power transfer, isolating one circuit from another, or isolating direct current (DC) while maintaining alternating current (AC) continuity between two circuits.

Essentially, a transformer consists of two or more windings coupled by mutual magnetic flux. If one of these windings, the primary, is connected to an alternating voltage source, an alternating flux will be produced whose amplitude will depend on the primary voltage, the frequency of the applied voltage, and the number of turns. The mutual flux will link the other winding, the secondary, and will induce a voltage in it whose value is depended on the number of secondary turns as well as the magnitude of the mutual flux and the frequency. By properly proportioning the number of primary and secondary turns, almost any desired voltage ratio, or ratio of transformation, can be obtained.

The essence of transformer action requires only the existence of time-varying mutual flux linking two windings. Such action can occur for two windings coupled through air, but coupling between the windings can be made much more effectively using a core of iron or other ferromagnetic material, because most of the flux is then confined to a definite, high-permeability path linking the windings. Such a transformer is commonly called an iron-core transformer. Most transformers are of this type. The following discussion is concerned almost wholly with iron-core transformers.

In order to reduce the losses caused by eddy currents in the core, the magnetic circuit usually consists of a stack of thin laminations. Two common types of construction are shown schematically in Fig. 1. 1. In the coretype [Fig. 1. 1 (a)] construction, the windings are wound around two legs of a rectangular magnetic core; in the shell-type [Fig. 1. 1 (b)] construction, the windings are wound around the center leg of a three legged core. Silicon-steel laminations of 0. 35 mm thickness are generally used for transformers operating at frequencies below a few hundred Hz. Silicon steel has the desirable properties of low cost, low core loss, and high permeability at high flux densities (1. 0 to 1. 5T). The cores of small transformers used in communication circuits at higher frequencies and lower energy levels are sometimes made of compressed powdered ferromagnetic alloys known as ferrites.

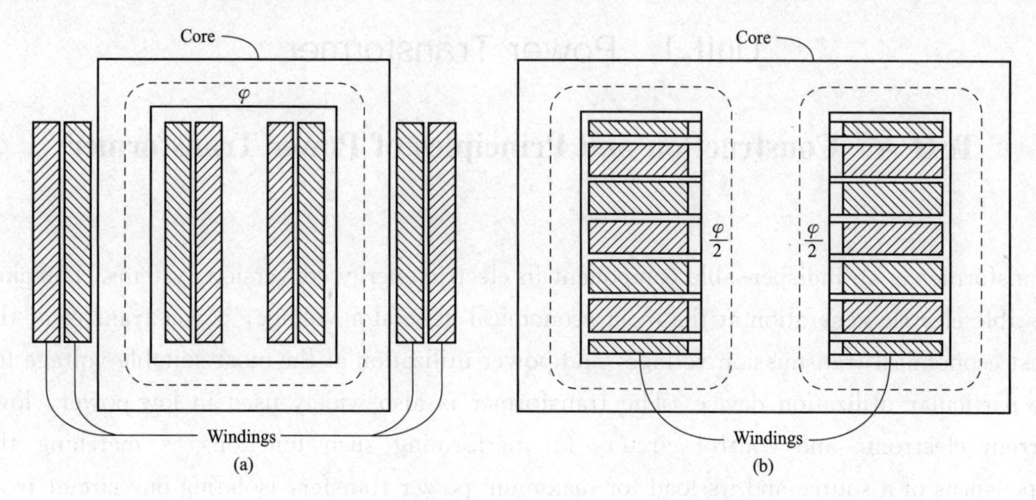

Fig. 1. 1 Schematic views of transformers

(a) core-type construction; (b) shell-type construction

In both configurations, most of the flux is confined to the core and linking both windings. The windings also produce an additional flux, known as leakage flux, which links one winding without linking the other. Although leakage flux is small fraction of the total flux, it plays an important role in determining the behavior of the transformer. In practical transformers, leakage is reduced by subdividing the windings into sections placed as close together as possible. In the core-type construction, each winding consists of two sections, one section on each of the two legs of the core, the primary and secondary windings being concentric coils. In the shell-type construction, variations of the concentric-winding arrangement may be used, or the windings may consist of a number of thin pancake coils assembled in a stack with primary and secondary coils interleaved.

Words and Phrases

power transformer 电力变压器
indispensable *adj*. 不可缺少的
power transfer 电能传输，功率转换
isolate *v*. 隔离，孤立，绝缘
matching the impedance 阻抗匹配
winding *n*. 绕组
construction *n*. 结构
primary *adj*. 一次测的
secondary *adj*. 二次测的
turn *n*. 匝数
time-varying *adj*. 时变的
stack *n*. 堆，叠
eddy current 涡流
lamination *n*. 薄层，叠片

silicon-steel laminations 硅钢片
couple *v.* 耦合
loss *n.* 损耗
leakage flux 漏磁通
core type 芯式
shell type 壳式
core *n.* 铁心
wound *v.* (wind 的过去分词) 缠，绕
powdered *adj.* 粉末状的，弄成粉的
ferromagnetic *adj.* 铁磁的
alloy *n.* 合金
permeability *n.* 磁导率
interleaved *adj.* 交叉放置的
concentric coils 同心式线圈
pancake coil 扁平线圈
legs of the core 铁心柱

Notes

1. Such action can occur for two windings coupled through air, but coupling between the windings can be made much more effectively using a core of iron or other ferromagnetic material, because most of the flux is then confined to a definite, high-permeability path linking the windings.

这样的作用也可以发生在通过空气耦合的两个绕组中，但用铁心或其他铁磁材料可以使绕组间的耦合作用更强，因为大部分磁通被限制在与两个绕组交链的高磁导率的路径中。

2. In the core type the windings are wound around two legs of a rectangular magnetic core [Fig. 1.1 (a)]; in the shell type the windings are wound around the center leg of a three-legged core [Fig. 1.1 (b)].

芯式变压器的绕组绕在两个矩形铁心柱上 [图 1.1 (a)]。壳式变压器的绕组绕在铁心柱的中间心柱上 [图 1.1 (b)]。

Exercises

1. Answer the following questions according to the text

(1) How many types of core construction do power transformers have?

(2) Why does power transformer consist of a stack of thin laminations?

(3) What does the alternating flux amplitude depend on?

(4) What is leakage flux?

(5) In order to reduce the losses caused by eddy currents in the core, Silicon-steel laminations are generally used for transformers operating at frequencies below a few hundred Hz. Right or wrong?

2. Translate the following sentences into Chinese according to the text

(1) The transformer is also widely used in low-power, low-current electronic and control circuits for performing such functions as matching the impedances of a source and its load for maximum power transfer.

(2) By properly proportioning the number of primary and secondary turns, almost any desired voltage ratio, or ratio of transformation, can be obtained.

(3) The windings also produce additional flux, known as leakage flux, which links one winding without linking the other.

(4) In the core-type construction, each winding consists of two sections, one section on each of the two legs of the core, the primary and secondary windings being concentric coils.

3. Translate the following paragraph into Chinese

In addition to the various power transformers, two special-purpose transformers are used with electric machinery and power systems. The first of these special transformers is a device specially designed to sample a high voltage and produce a low secondary voltage directly proportional to it. Such a transformer is a potential transformer (电压互感器). A power transformer also produces a secondary voltage directly proportional to its primary voltage; the difference between a potential transformer and a power transformer is that the potential transformer is designed to handle only a very small current. The second type of special transformer is a device designed to provide a secondary current much smaller than but directly proportional to its primary current. This device is called a current transformer (电流互感器).

Text B Differences between Transformers and Rotating Machines

In both transformers and rotating machines, a magnetic field is created by the combined action of the currents in the windings. In an iron-core transformer, most of this flux is confined to the core and links all the windings. This resultant mutual flux induces voltages in the windings proportional to their number of turns and is responsible for the voltage-changing property of a transformer. In rotating machines, the situation is similar, although there is an air gap which separates the rotating and stationary components of the machine. Directly analogous to the manner in which transformer core flux links the various windings on a transformer core, the mutual flux in rotating machines crosses the air gap, linking the windings on the rotor and stator. As in a transformer, the mutual flux induces voltages in these windings proportional to the number of turns and the time rate of change of the flux.

A significant difference between transformers and rotating machines is that in rotating machines there is relative motion between the windings on the rotor and stator. This relative motion produces an additional component of the time rate of change of the various winding flux linkages. The resultant voltage component, known as the speed voltage, is characteristic of the process of electromechanical energy conversion. In a static transformer, however, the time variation of flux linkages is caused simply by the time variation of winding currents; no mechanical motion is involved, and no electromechanical energy conversion takes place.

The resultant core flux in a transformer induces a counterElectro-Motive Force (EMF) in the primary which, together with the primary resistance and leakage-reactance voltage drops, must balance the applied voltage. Since the resistance and leakage-reactance voltage drops usually are small, the counter EMF must approximately equal to the applied voltage and the core flux must adjust itself accordingly. Exactly similar phenomena must take place in the armature windings of an AC motor; the resultant air-gap flux wave must adjust itself to generate a counter EMF approximately equal to the applied voltage. In both transformers and rotating machines, the Magneto-Motive Force (MMF) of all the currents must accordingly adjust itself to create the resultant flux required by this voltage balance. In any AC electromagnetic device in which the resistance and leakage-reactance voltage drops are small, the resultant flux is very nearly determined by the applied voltage and frequency, and the currents must adjust themselves accordingly to produce the MMF required to create this flux.

In a transformer, the secondary current is determined by the voltage induced in the secondary winding, the secondary leakage impedance, and the electric load. In an induction motor, the secondary (rotor) current is determined by the voltage induced in the secondary, the secondary leakage impedance, and the mechanical load on its shaft. Essentially the same phenomena place in the primary winding of the transformer and in the armature (stator) windings of induction and synchronous motors. In all three, the primary, or armature, current must adjust itself so that the combined MMF of all currents creates the flux required by the applied voltage.

In addition to the useful mutual fluxes, in both transformers and rotating machines have leakage fluxes that link individual windings without linking others. Although the detailed picture of the leakage fluxes in rotating machines is more complicated than that in transformers, their effects are essentially the same. In both, the leakage fluxes induce voltages in AC windings which are accounted for as leakage-reactance voltage drops. In both, the reluctances of the leakage-flux paths are dominated by that of a path through air, and hence the leakage fluxes are nearly linearly proportional to the currents producing them. The leakage-reactance therefore is often assumed to be constant, independent of the degree of saturation of the main magnetic circuit.

Further examples of the basic similarities between transformers and rotating machines can be cited. Except for friction and windage, the losses in transformers and rotating machines are essentially the same. Tests for determining the losses and equivalent circuit parameters are similar: an open-circuit, or no-load, test gives information regarding the excitation requirements and core losses (along with friction and windage losses in rotating machines), while a short-circuit test together with DC resistance measurements gives information regarding leakage reactance and winding resistance.

Words and Phrases

rotating machine 旋转电机

combined action 共同作用

resultant *adj*. 合成的

link all the windings 与所有绕组交链

mutual *adj*. 共有的，相互的

gap *n*. 气隙

rotor *n*. 转子
stator *n*. 定子
proportional *adj*. 成正比的
relative motion 相对运动
counter *adj*. 相反的
EMF *n*. 电动势
voltage drop 电压降
MMF *n*. 磁动势
AC *abbr*. 交流
shaft *n*. 轴
leakage-reactance *n*. 漏电抗
constant *n*. 常数
time rate of change of the flux 磁通随时间的变化率
flux linkage 磁通链，磁链
synchronous motor 同步电动机
friction losses 摩擦损耗
windage losses 风耗
core losses 铁心损耗
no-load test 空载实验
short-circuit test 短路实验
regarding leakage reactance and winding resistance 考虑漏电抗和绕组电阻

Notes

1. This resultant mutual flux induces voltages in the windings proportional to their number of turns and is responsible for the voltage-changing property of a transformer.

变压器合成磁通在绕组中感应出的电压与绕组匝数呈正比，并且变压器电压变化特性取决于合成磁通。

2. Directly analogous to the manner in which transformer core flux links the various windings on a transformer core，the mutual flux in rotating machines crosses the air gap，linking the windings on the rotor and stator.

变压器铁心磁通交链铁心柱上的不同绕组，旋转电机的主磁通穿过气隙与定子绕组和转子绕组交链。

3. Tests for determining the losses and equivalent circuit parameters are similar：an open-circuit，or no-load，test gives information regarding the excitation requirements and core losses.

确定各种损耗和等效电路参数的测试是相同的：通过开路或空载实验可以得到励磁参数和铁心损耗。

Exercises

1. Answer the following questions according to the text

(1) What is the significant difference between transformers and rotating machines?

(2) What determines the secondary current in a transformer?

(3) In the fourth paragraph，there is a sentence：In all three，the primary，or armature，current must adjust itself so that the combined MMF of all currents creates the flux required by the applied voltage. What does the three indicates.

(4) The leakage-reactance is often assumed to be constant，independent of the degree of

saturation of the main magnetic circuit. Why?

(5) What tests are used for determining the losses and equivalent circuit parameters of transformers and rotating machines?

2. Translate the following sentences into Chinese according to the text

(1) The resultant core flux in a transformer induces a counter Electro-Motive Force (EMF) in the primary which, together with the primary resistance and leakage-reactance voltage drops, must balance the applied voltage.

(2) In any AC electromagnetic device in which the resistance and leakage-reactance voltage drops are small, the resultant flux is very nearly determined by the applied voltage and frequency, and the currents must adjust themselves accordingly to produce the MMF required to create this flux.

(3) In addition to the useful mutual fluxes, in both transformers and rotating machines there are leakage fluxes which link individual windings without linking others.

(4) In addition to the useful mutual fluxes, in both transformers and rotating machines there are leakage fluxes which link individual windings without linking others.

3. Translate the following paragraph into Chinese

In electronic power supplies there is often a need to isolate the output from the input and to reduce the weight and cost of the unit. In other applications, such as in aircraft, there is a strong incentive to minimize weight. These objectives are best achieved by using a relatively high frequency transformer. Thus, in aircraft the frequency of transformers is typically 400 Hz, while in electronic power supplies the frequency of transformers may range from 5 kHz to 50 kHz.

Unit 2 Direct Current (DC) Machines

Text A Direct Current Machines

Commercial DC generators and motors are built the same way; consequently, any DC generator can operate as a motor and vice versa. The armature winding of a DC generator is on the rotor with current conducted from it by means of carbon brushes. The field winding is on the stator and is excited by direct current.

The armature winding, consisting of a single coil of *N* turns, is indicated by the two coil sides *a* and -*a* placed at diametrically opposite points on the rotor with the conductors parallel to the shaft. See Fig. 1. 2. The rotor is normally turned at a constant speed by a source of mechanical power connected to the shaft. The air-gap flux distribution usually approximates a flat-topped wave, rather than the sine wave found in AC machines. Rotation of the coil generates a coil voltage which is a time function having the same waveform as the spatial flux-density distribution.

Although the ultimate purpose is the generation of a direct voltage, the voltage induced in an individual armature coil is an alternating voltage, which must therefore be rectified. The output voltage of an AC machine can be rectified using external semiconductor rectifiers. This is in contrast to the conventional DC machine in which rectification is produced mechanically by means of a commutator, which is a cylinder formed of copper segments insulated from each other by mica or some other highly insulating material and mounted on, but insulated from, the rotor shaft. Stationary carbon brushes held against the commutator surface connect the winding to the external armature terminals. The need for commutation is the reason why the armature windings of DC machines are placed on the rotor.

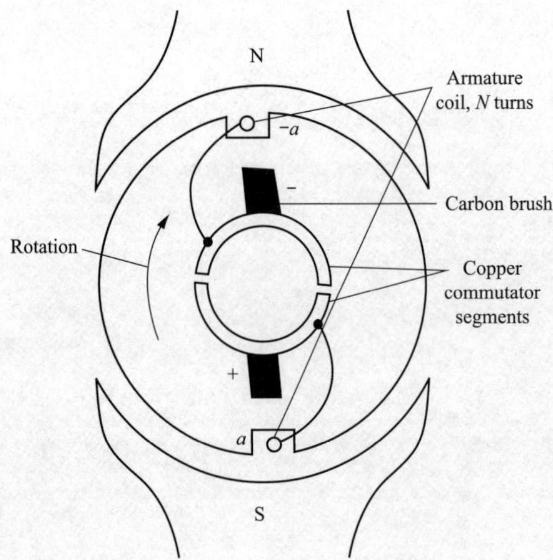

Fig. 1. 2 Schematic diagram of DC machine with commutator

The commutator at all time connects the coil side, which is under the South Pole, to the positive brush and that under the North Pole to the negative brush in Fig. 1. 2. The commutator provides full-wave rectification, transforming the voltage waveform between brushes to that of Fig. 1. 3 (b) and making an available unidirectional voltage to the external circuit. By increasing the number of coils and segments, which can decrease the pulsation of the DC voltage, we can obtain a DC voltage that is very smooth. Modern DC generators produce voltages having a ripple of less than 5 percent. The DC machine of Fig. 1. 2 is, of course, simplified to the point of being unrealistic in the practical sense, but the operating principle is clear to be understood.

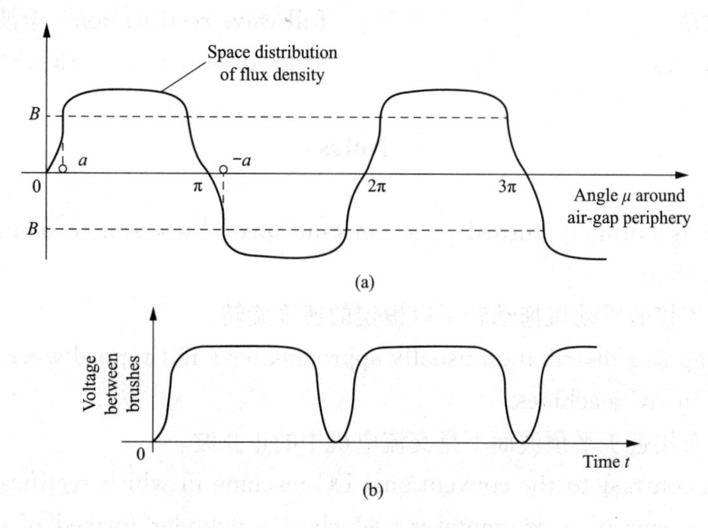

Fig. 1. 3 Air gap flux density and voltage

(a) Space distribution of air-gap flux density in an elementary DC machine;

(b) Waveform of voltage between brushes

The effect of direct current in the field winding of a DC machine is to create a magnetic flux distribution which is stationary with respect to the stator. Similarly, the effect of the commutator is that when direct current flows through the brushes, armature creates a magnetic flux distribution which is also fixed in space and whose axis, determined by the design of the machine and the position of the brushes, is typically perpendicular to the axis of the field flux.

Thus, just as in the AC machines, it is the interaction of these two flux distributions that creates the torque of the DC machine. If the machine is acting as a generator, this torque opposes rotation. If it is acting as a motor, the electromechanical torque acts in the direction of the rotation.

Words and Phrases

Direct Current (DC) 直流
generator *n.* 发电机
motor *n.* 电动机
armature *n.* 电枢
armature winding 电枢绕组
excite *v.* 激励
field winding 励磁绕组
flat-topped wave 平顶波
flux-density *n.* 磁通密度
rectify *v.* 调整，整定
rectifier *n.* 整流器
semiconductor rectifier 半导体整流器
commutator *n.* 换向器
mica *n.* 云母

insulating material 绝缘材料
carbon brush 碳刷（电刷）
unidirectional *n.* 单方向的
pulsation *n.* 脉动
ripple *n.* 波动，波纹
torque *n.* 转矩
periphery *n.* 外围
in contrast to 和……成对照
full-wave rectification 全波整流
space distribution 空间分布

Notes

1. The rotor is normally turned at a constant speed by a source of mechanical power connected to the shaft.

与转子轴相连接的原动机拖动转子以恒定的速度旋转。

2. The air-gap flux distribution usually approximates a flat-topped wave，rather than the sine wave found in AC machines.

气隙磁场分布接近于平顶波而不是交流电机中的正弦波。

3. This is in contrast to the conventional DC machine in which rectification is produced mechanically by means of a commutator，which is a cylinder formed of copper segments insulated from each other by mica or some other highly insulating material and mounted on，but insulated from，the rotor shaft.

这是与传统直流电机的一个对照，传统电机用换向器来完成机械整流，换向器是安装在转子轴上且与轴绝缘的由很多铜片组成的圆柱体，换向片与换向片之间用云母或其他高绝缘材料绝缘。

4. Similarly，the effect of the commutator is that when direct current flows through the brushes，armature creates a magnetic flux distribution which is also fixed in space and whose axis，determined by the design of the machine and the position of the brushes，is typically perpendicular to the axis of the field flux.

同样，换向器的作用是当直流电流通过电刷、电枢时，电枢电流会产生一个空间固定的磁通分布，此磁通的方向由电机结构和电刷位置决定，典型方向是与主极磁场的磁通方向正交。

Exercises

1. Answer the following questions according to the text

(1) A DC generator can operate as a DC motor and vice versa. Right or wrong?

(2) The air-gap flux distribution in DC electric machine usually approximates a sine wave. Right or wrong?

(3) The commutator of DC electric machine provides full-wave rectification. Right or wrong?

(4) How to decrease the pulsation of the DC voltage?

(5) How to create the torque of the DC machine?

2. Translate the following sentences into Chinese according to the text

(1) Although the ultimate purpose is the generation of a direct voltage, the voltage induced in an individual armature coil is an alternating voltage, which must therefore be rectified.

(2) The need for commutation is the reason why the armature windings of DC machines are placed on the rotor.

(3) The commutator at all times connects the coil side, which is under the South Pole, to the positive brush and that under the North Pole to the negative brush.

(4) The effect of direct current in the field winding of a DC machine is to create a magnetic flux distribution which is stationary with respect to the stator.

(5) If the machine is acting as a generator, this torque opposes rotation. If it is acting as a motor, the electromechanical torque acts in the direction of the rotation.

3. Translate the following paragraph into Chinese

In general, the outstanding advantage of DC machines lies in their flexibility and versatility. Before the widespread availability of AC motor drives, DC machines were essentially the only choice available for many applications requiring a high degree of control. Their principal disadvantages stem from the complexity associated with the armature winding and the commutator/brush system. Not only does this additional complexity increase the cost over competing AC machines, it also increases the need for maintenance and reduces the potential reliability of these machines.

Text B Elementary Knowledge of Rotating Machines

Electromagnetic energy conversion occurs when changes in the flux linkage result from mechanical motion. In rotating machines, voltages are generated in windings or groups of coils by rotating these windings mechanically through a magnetic field, by mechanically rotating a magnetic field past the winding, or by designing the magnetic circuit so that the reluctance varies with rotation of the rotor. By any of these methods, the flux linking a specific coil is changed cyclically, and a time-varying voltage is generated.

A set of such coils connected together is typically referred to as an armature winding. In general, the term armature winding is used to refer to a winding or a set of windings on a rotating machine which carry AC currents. In AC machines such as synchronous or induction machines, the armature winding is typically on the stationary portion of the motor referred to as the stator, in which case these windings may also be referred to as stator windings.

In a DC machine, the armature winding is found on the rotating member, referred to as the rotor. The armature winding of a DC machine consists of many coils connected together to form a closed loop. A rotating mechanical contact is used to supply current to the armature winding as the rotor rotates.

Synchronous and DC machines typically include a second winding (or set of windings) which carry DC current and which are used to produce the main operating flux in the machine. Such a winding is typically referred to as field winding. The field winding on a DC machine is found on the stator, while that on a synchronous machine is found on the rotor, in which case current must be supplied to the field winding via a rotating mechanical contact. As we have seen, permanent magnets also produce DC magnetic flux and are used in the place of field windings in some machines.

In most rotating machines, the stator and rotor are made of electrical steel, and the windings are installed in slots on these structures. The use of such high-permeability material maximizes the coupling between the coils and increases the magnetic energy density associated with the interaction. It also enables the machine designer to shape and distribute the magnetic fields according to the requirements of each particular machine design. The time varying flux present in the armature structures of these machines tends to induce currents, known as eddy currents, in the electrical steel. Eddy currents can be a large source of loss in such machines and can significantly reduce machine performance. In order to minimize the effects of eddy currents, the armature structure is typically built from thin laminations of electrical steel which are insulated from each other.

In some machines, such as variable reluctance machines and stepper motors, there are no windings on the rotor. Operation of these machines depends on the nonuniformity of air-gap reluctance associated with variations in rotor position in conjunction with time-varying currents applied to their stator windings. In such machines, both the stator and rotor structures are subjected to time-varying magnetic flux and, as a result, both may require lamination to reduce eddy-current losses.

Rotating electric machines take many forms and are known by many names: DC, synchronous, permanent-magnet, induction, variable reluctance, hysteresis, brushless, and so on. Although these machines appear to be quite dissimilar, the physical principles governing their behavior are quite similar, and it is often helpful to think of them in terms of the same physical picture.

Words and Phrases

mechanical *adj*. 机械的
motion *n*. 运动
reluctance *n*. 磁阻
cyclically *adv*. 周期性地，循环地
term *n*. 术语
referred to as 叫作，称为
synchronous machine 同步电机
induction machine 感应电机
permanent magnet 永磁体，永久磁铁
slot *n*. 槽
high-permeability *adj*. 高磁导率的
insulate *v*. 绝缘
nonuniformity *n*. 不均匀性
brushless *n*. 无刷
electrical steel 电工钢
magnetic energy density 磁能密度
variable reluctance machine 变磁阻电机
stepper motor 步进电动机

air-gap reluctance 气隙磁阻 rotor position 转子位置

Notes

1. In rotating machines, voltages are generated in windings or groups of coils by rotating these windings mechanically through a magnetic field, by mechanically rotating a magnetic field past the winding, or by designing the magnetic circuit so that the reluctance varies with rotation of the rotor.

通过线圈在磁场中旋转，或者绕组中的磁场旋转，或者通过设计磁路使磁阻随转子的转动而变化，都可以使旋转电机中绕组或线圈组中产生电压。

2. In AC machines such as synchronous or induction machines, the armature winding is typically on the stationary portion of the motor referred to as the stator, in which case these windings may also be referred to as stator windings.

在交流电机如同步电机或感应电机中，电枢绕组一般安放在电机的固定部分即定子上，这些绕组也可以叫作定子绕组。

Exercises

1. Answer the following questions according to the text

(1) All Rotating machines have a rotating rotor. Right or wrong?

(2) What is the name of the windings which carry DC current and are used to produce the main operating flux in the machine?

(3) Permanent magnets can also produce DC magnetic flux in rotating machines. Right or wrong?

(4) Why is the armature structure built into thin laminations of electrical steel which are insulated from each other?

(5) How many names do you know of rotating electric machines? What are they?

2. Translate the following sentences into Chinese according to the text

(1) The flux linking a specific coil is changed cyclically, and a time-varying voltage is generated.

(2) In general, the term armature winding is used to refer to a winding or a set of windings on a rotating machine which carry AC currents.

(3) In a DC machine, the armature winding is found on the rotating member, referred to as the rotor. The armature winding of a DC machine consists of many coils connected together to form a closed loop.

(4) The use of such high-permeability material maximizes the coupling between the coils and increases the magnetic energy density associated with the interaction.

(5) Rotating electric machines take many forms and are known by many names: DC, synchronous, permanent-magnet, induction, variable reluctance, hysteresis, brushless, and

so on.

3. Translate the following paragraph into Chinese

There are three basic type of rotating machines: synchronous, induction, and DC machines. In all of them the basic principles are essentially the same. Voltages are generated by the relative motion of a magnetic field with respect to a winding, and torques are produced by the interaction of the magnetic fields of the stator and rotor windings.

Unit 3 Alternating Current (AC) Machines

Text A Synchronous Machines

A preliminary picture of synchronous machine performance can be gained by discussing the voltage induced in the armature of the very much simplified salient-pole AC synchronous generator shown in Fig. 1. 4. The field-winding of this machine produces a single pair of magnetic poles, hence this machine is referred to as a two-pole machine.

The armature winding of a synchronous machine is on the stator, and the field winding is on the rotor. The field winding is excited by direct current conducted to it by means of stationary carbon brushes which contact rotating slip rings or collector rings. Practical factors usually dictate this orientation of the two windings: It is advantageous to have the single, low-power field winding on the rotor while having the high-power, typically multiple-phase, armature winding on the stator.

The armature winding, consisting here of only a single coil of N turns, is indicated in cross section by the two coil sides a and $-$a placed in diametrically opposite narrow slots on the inner periphery of the stator of Fig. 1. 4. The conductors forming these coil sides are parallel to the shaft of the machine and are connected in series (not shown in the figure). The rotor is turned at a constant speed by a source of mechanical power connected to its shaft. The armature winding is assumed to be open-circuited and hence the flux in this machine is produced by the field winding alone. Flux paths are shown schematically by dashed lines in Fig. 1. 4.

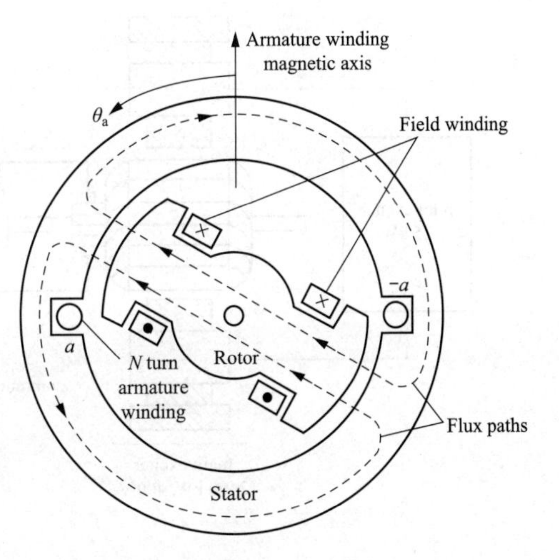

Fig. 1. 4 Schematic view of a simple, two-pole, single-phase synchronous generator

A highly idealized analysis of this machine would assume a sinusoidal distribution of magnetic flux in the air gap. The resultant space distribution of air-gap flux density B is shown in Fig. 1. 5 (a) as a function of the spatial angle θ_a around the rotor periphery. In practice, the air-gap flux-density of practical salient-pole machines can be made to approximate a sinusoidal distribution by properly shaping the pole faces.

As the rotor rotates, the flux-linkages of the armature winding change with time. Under the assumption of a sinusoidal flux distribution and constant rotor speed, the resulting coil voltage will be sinusoidal in time as shown in Fig. 1. 5 (b). The coil voltage passes through a complete cycle for each revolution of the two-pole machine of Fig. 1. 4. Its frequency (Hz) in cycles per second is the same as the speed of the rotor in revolutions per second: the electric frequency of the generated voltage is synchronized with the mechanical speed, and this is the reason for the designation "synchronous" machine. Thus, a two-pole synchronous machine must revolve at 3000 revolutions per minute to produce a 50 Hz voltage.

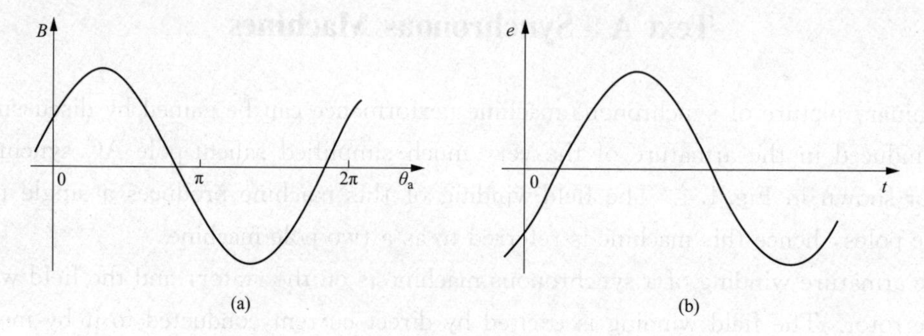

Fig. 1. 5 Air gap flux density and voltage

(a) Space distribution of flux density; (b) corresponding waveform of the generated voltage for the single-phase generator of Fig. 1. 4

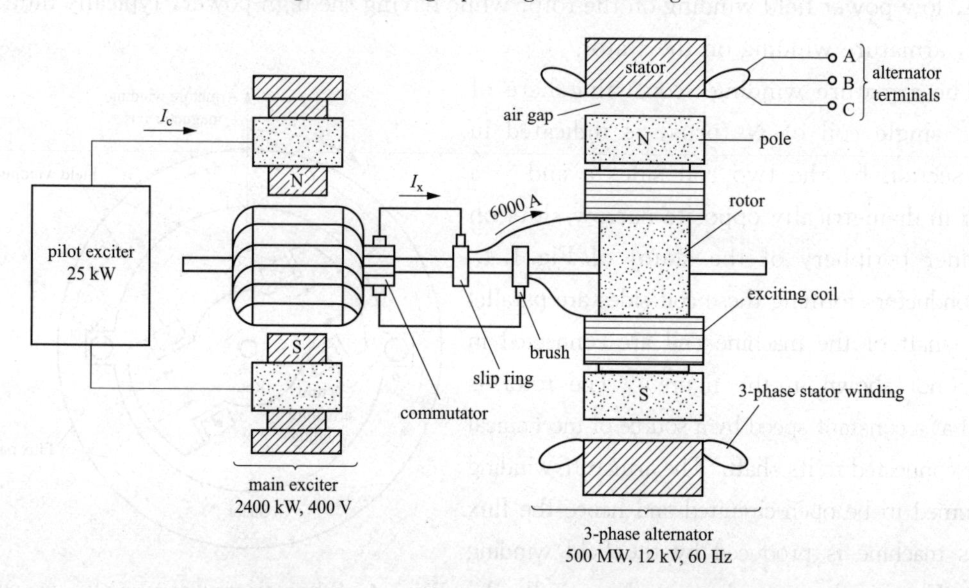

Fig. 1. 6 Schematic diagram and cross-section view of a typical 500 MW synchronous generator and its 2400 kW DC exciter.

The DC exciting current I_x (6000 A) flows though the commutator and two slip-rings. The DC control current I_c from the pilot exciter permits variable field control of the main exciter, which in turn controls I_x.

A great many synchronous machines have more than two poles. The 3-phase stator winding of revolving-field synchronous generator is directly connected to the load, without going through large, unreliable slip-rings and brushes. A stationary stator also makes it easier to insulate the windings because they are not subjected to centrifugal forces. Fig. 1. 6 is a schematic diagram of such a generator, sometimes called an alternator. The field is excited by a DC generator, usually mounted on the same shaft. Note that the brushes on the commutator have to be connected to another set of brushes riding on slip-rings to feed the DC current I_x into the revolving field.

The counterpart of the synchronous generator is the synchronous motor. To produce a steady electromechanical torque, the magnetic fields of the stator and rotor must be constant in amplitude and stationary with respect to each other. In a synchronous motor, the steady-state speed is determined by the number of poles and the frequency of the armature current.

In both generators and motors, an electromechanical torque and a rotational voltage are produced. These are the essential phenomena for electromechanically energy conversion.

Words and Phrases

alternating current 交流
salient-pole *n.* 凸极
collector ring 集电环
open-circuit *v.* 开路
revolution *n.* 转速，转数
wavelength *n.* 波长
magnetic pole 磁极
cross section 横截面
coil side 线圈边
inner periphery of the stator 定子内圆表面
connected in series 串联
schematic view 示意图
flux path 磁路
dashed line 虚线
sinusoidal distribution 正弦分布
flux density 磁通密度
pole face 极面
magnetic flux in the air gap 气隙磁通
revolving-field *adj.* 旋转磁场的
terminal *n.* 终端，接线端
centrifugal force 离心力
main exciter 主励磁机
pilot exciter 副励磁机
rotational *adj.* 转动的，轮流的
counterpart *n.* 配对物

Notes

1. The conductors forming these coil sides are parallel to the shaft of the machine and are connected in series (not shown in the figure). The rotor is turned at a constant speed by a source of mechanical power connected to its shaft.

组成线圈边的导体与电机的轴平行并互相串联（图中没有显示出），转子由与它的轴连接的原动机拖动并以恒定的速度旋转。

2. The coil voltage passes through a complete cycle for each revolution of the two-pole machine of Fig. 1. 4. Its frequency in cycles per second (Hz) is the same as the speed of the rotor in revolutions per second: the electric frequency of the generated voltage is synchronized with the mechanical speed, and this is the reason for the designation "synchronous" machine.

图 1. 4 所示的两极电机每转一圈，线圈电压就经过一个完整周期，电压的交变频率等于用每秒转数为单位的转子转速，即感应电压的电气频率与机械转速同步，这也是同步电机的设计根据。

Exercises

1. Answer the following questions according to the text

(1) When the armature winding of synchronous machine is assumed to be open-circuited, the flux in this machine is produced by the field winding alone. Right or wrong?

(2) What is the reason for the designation "synchronous" machine?

(3) How many revolutions per minute must revolve at for a two-pole synchronous machine to produce a 60 Hz voltage?

(4) All of synchronous machines have two poles. Right or wrong?

(5) What determines the steady-state speed of a synchronous motor?

2. Translate the following sentences into Chinese according to the text

(1) The field-winding of this machine produces a single pair of magnetic poles, and hence this machine is referred to as a two-pole machine.

(2) In practice, the air-gap flux-density of practical salient-pole machines can be made to approximate a sinusoidal distribution by properly shaping the pole faces.

(3) The brushes on the commutator have to be connected to another set of brushes riding on slip-rings to feed the DC current I_x into the revolving field.

(4) The counterpart of the synchronous generator is the synchronous motor.

(5) In both generators and motors, an electromechanical torque and a rotational voltage are produced. These are the essential phenomena for electromechanically energy conversion.

3. Translate the following paragraph into Chinese

Synchronous machines can be classified as cylindrical-rotor (隐极式转子) or salient-pole (凸极式) machines. The cylindrical-rotor construction is used in high-speed steam-turbine-driven generators (汽轮发电机). The armature windings consist of laminated conductors placed in the stator slots. The rotor carries the DC field winding. Most of turbine generators being built at present for 60 Hz service are 2-pole 3600 revolutions per minute machines. Because of the economies of high-speed high-temperature high-pressure steam turbines, much study and some real pioneering work have been devoted to improvements in materials and design of both generators and turbines.

Text B Induction Machines

A second type of AC machine is induction machine. Like the synchronous machine, the stator winding of an induction machine is excited with alternating currents. In contrast to a synchronous machine in which a field winding on the rotor is excited with DC current, alternating currents flow in the rotor windings of an induction machine. In induction machines, alternating currents are applied directly to the stator windings. Rotor currents are then produced by induction, i. e. transformer action. The induction machine may be regarded as a generalized transformer in which electric power is transformed between rotor and stator together with a change of frequency and a flow of mechanical power. Although the induction motor is the most common of all motors, it is seldom used as a generator. Its performance characteristics as a generator are unsatisfactory for most applications, although in recent years it has been found to be well suited for wind-power applications. The induction machine may also be used as a frequency changer.

In the induction motor, the stator windings are essentially the same as those of a synchronous machine. However, the rotor windings are electrically short-circuited and frequently have no external connections; currents are induced by transformer action from the stator winding. The rotor "windings" of a squirrel cage induction motor are actually solid aluminum bars which are cast into the slots in the rotor and which are shorted together by cast aluminum rings at each end of the rotor. This type of rotor construction results in induction motors which are relatively inexpensive and highly reliable, factors contributing to their immense popularity and widespread application.

As in a synchronous motor, the armature flux in the induction motor leads that of the rotor and produces an electromechanical torque. In fact, we will see that, just as in a synchronous machine, the rotor and stator fluxes rotate in synchronism with each other and that torque is related to the relative displacement between them. However, unlike a synchronous machine, the rotor of an induction machine does not itself rotate synchronously; it is the "slipping" of the rotor with respect to the synchronous armature flux that gives rise to the induced rotor currents and hence the torque. Induction motors operate at speeds less than the synchronous mechanical speed. A typical speed-torque characteristic for an induction motor is shown in Fig. 1. 7.

Three phase induction motors are the motors most frequently encountered in industry. They are simple, low-priced, and easy to maintain. They run at essentially constant speed from zero to full-load. The speed is frequency-dependent and, consequently, these motors are not easily adapted to speed control. However, variable frequency electronic drives are being used more and more to control the speed of commercial induction motors.

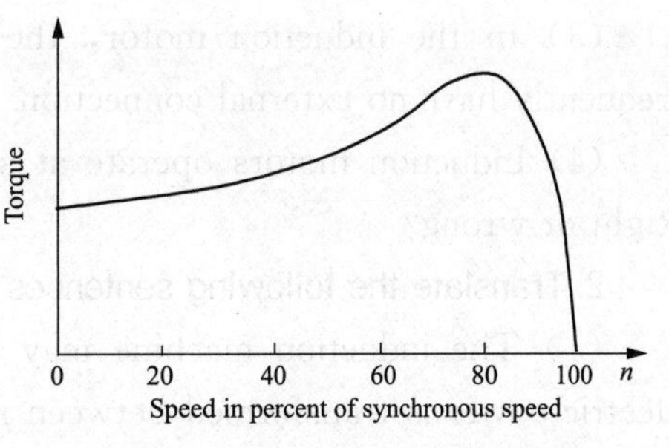

Fig. 1. 7 Typical induction-motor speed-torque characteristic

Words and Phrases

electric power 电功率，电力

generalized transformer 广义变压器

mechanical power 机械功率

seldom *adv.* 很少

wind-power *n.* 风力

short-circuit *n.* 短路

squirrel cage *n.* 笼型

give rise to 引起，使产生

encounter *v.* 遇到
maintain *v.* 维修，维护
full-load *n.* 满载
frequency changer 频率变换器
immense popularity 非常普及
variable frequency electronic drive 变频电子装置

Notes

1. In contrast to a synchronous machine in which a field winding on the rotor is excited with DC current, alternating currents flow in the rotor windings of an induction machine.

同步电机的转子上是励磁绕组并且是直流励磁，而感应电机的转子绕组中流过的是交流电流。

2. The rotor "windings" of a squirrel-cage induction motor are actually solid aluminum bars which are cast into the slots in the rotor and which are shorted together by cast aluminum rings at each end of the rotor.

笼型感应电机的转子绕组通常是浇铸在转子槽中的固体铝条，这些铝条被转子两头的浇铸铝环短接。

Exercises

1. Answer the following questions according to the text

(1) There are two typical AC electric machine, what are they?

(2) The currents flow in the rotor windings of an induction machine are alternating currents or direct currents?

(3) In the induction motor, the rotor windings are electrically short-circuited and frequently have no external connection. Right or wrong?

(4) Induction motors operate at speeds less than the synchronous mechanical speed. Right or wrong?

2. Translate the following sentences into Chinese according to the text

(1) The induction machine may be regarded as a generalized transformer in which electric power is transformed between rotor and stator together with a change of frequency and a flow of mechanical power.

(2) This type of rotor construction results in induction motors which are relatively inexpensive and highly reliable, factors contributing to their immense popularity and widespread application.

(3) Unlike a synchronous machine, the rotor of an induction machine does not itself rotate synchronously; it is the "slipping" of the rotor with respect to the synchronous armature flux that gives rise to the induced rotor currents and hence the torque.

(4) Three phase induction motors are the motors most frequently encountered in industry. They are simple, low-priced, and easy to maintain.

(5) More and more variable frequency electronic drives are being used to control the speed of commercial induction motors.

3. Translate the following paragraph into Chinese

Rotating electric machines take many forms and are known by many names: DC, synchronous, permanent-magnet, induction, etc. Although these machines appear to be quite dissimilar and require a variety of analytical techniques, the physical principles governing their behavior are quite similar, and in fact these machines can often be explained from the same physical picture. An induction machine, in spite of many fundamental differences, works on exactly the same principle; one can identify flux distributions associated with the rotor and stator, which rotate in synchronism and which are separated by some torque-producing angular displacement.

Unit 4 Permanent Magnet (PM) Machines

Text A Introduction to Permanent Magnet (PM) Machines

1. Introduction

With the emergence of modern ferrite and rare earth magnets, permanent magnet (PM) machines are being increasingly used in applications where efficiency, or more importantly, losses are of concern. One such application is in servo drives in which the motor is commanded to operate for a long period at, or very near, zero speed. In such applications, the heat rejection of a conventional squirrel cage induction motor drive could be insufficient to keep the machine (usually the rotor) at tolerable levels. Since PM machines use magnets to supply the excitation current (current required to magnetize the iron parts), a substantial reduction of the stator current is possible, thereby greatly reducing the losses. Increased use of such machines will inevitably also become important in conventional operation from the utility supply when the cost of energy increases to a sufficiently high value.

The use of permanent magnets (PMs) in construction of electrical machines brings the following benefits:

- No electrical energy is absorbed by the field excitation system and thus there are no excitation losses which means substantial increase in efficiency.
- Higher power density and/or torque density than when using electromagnetic excitation.
- Better dynamic performance than motors with electromagnetic excitation (higher magnetic flux density in the air gap).
- Simplification of construction and maintenance.
- Reduction of prices for some types of machines.

2. Classification of permanent magnet electric motors

In general, rotary PM motors for continuous operation are classified into DC brush commutator motors, DC brushless motors and AC synchronous motors. The construction of a PM DC commutator motor is similar to a DC motor with the electromagnetic excitation system replaced by PMs. PM DC brushless and AC synchronous motor designs are practically the same: with a polyphaser stator and PMs located on the rotor. The only difference is in the control and shape of the excitation voltage: an AC synchronous motor is fed with more or less sinusoidal waveforms which in turn produce a rotating magnetic field. In PM DC brushless motors the armature current has a shape of a square (trapezoidal) waveform, only two phase windings (for Y connection) conduct the current at the same time, and the switching pattern is synchronized with the rotor angular position (electronic commutation).

The armature current of synchronous and DC brushless motors is not transmitted through brushes, which are subject to wear and require maintenance. Another advantage of the brushless motor is the fact that the power losses occur in the stator, where heat transfer conditions are good. Consequently, the power density can be increased as compared with a DC commutator motor. In addition, considerable improvements in dynamics can be achieved because the air gap magnetic flux density is high, the rotor has a lower inertia and there are no speed-dependent current limitations. Thus, the volume of a brushless PM motor can be reduced by 40% to 50% while still keeping the same rating as that of a PM commutator motor (Fig. 1. 8).

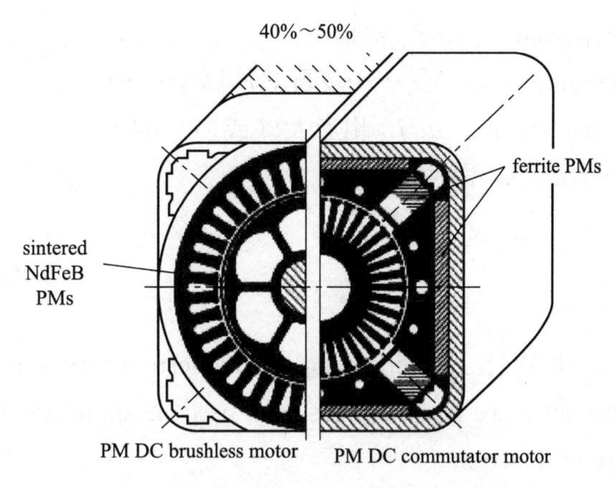

Fig. 1. 8 Comparison of PM brushless and PM DC commutator motors.

3. Applications of permanent magnet motors

PM motors are used in a broad power range from mW to hundreds kW. There are also attempts to apply PMs to large motors rated at minimum 1 MW. Thus, PM motors cover a wide variety of application fields, from stepping motors for wrist watches, through industrial drives for machine tools to large PM synchronous motors for ship propulsion (navy frigates, cruise ships, medium size cargo vessels and ice breakers). The automotive industry is the biggest user of PM DC commutator motors. The number of auxiliary DC PM commutator motors can vary from a few in an inexpensive car to about one hundred in a luxury car. PM brushless motors rated from 50 to 100 kW seem to be the best propulsion motors for electric and hybrid road vehicles.

Words and Phrases

ferrite *n.* 铁氧体

rare earth magnet 稀土磁体

servo drive 伺服驱动系统

squirrel cage induction motor drive 笼型感应电机驱动系统

tolerable *adj.* 可忍受的；过得去的

magnetize *adj.* 使……磁化；使……有磁性

substantial *adj.* 大量的；结实的，牢固的；重大的

inevitably *adv.* 不可避免地，自然而然地；必然地，无疑的

conventional *adj.* 传统的；习用的，平常的；依照惯例的；约定的

utility *n.* 功用，效用；有用的物体或器械；*adj.* 有多种用途的；各种工作都会做的

power/ torque density 功率、转矩密度

electromagnetic *adj.* 电磁的

dynamic performance 动态性能

flux *n.* （磁）通量

rotary *adj.* 旋转的

DC brush commutator motors 直流有刷换向电动机

DC brushless motors 直流无刷电动机

AC synchronous motors 交流同步电动机

sinusoidal *adj.* 正弦的

trapezoidal　*adj.* 梯形的
be subject to　受……支配；从属于……
considerable　*adj.* 相当大（或多）的
inertia　*n.* 惯性；惰性
auxiliary　*adj.* 辅助的；备用的，补充的；附加的；副的
hybrid　*adj.* 混合的

Notes

1. With the emergence of modern ferrite and rare earth magnets, permanent magnet (PM) machines are being increasingly used in applications where efficiency, or more importantly, losses are of concern.

随着现代铁氧体和稀土磁体的出现，永磁电机正越来越多地应用于更关注效率或损耗的场合中。

2. Increased use of such machines will inevitably also become important in conventional operation from the utility supply when the cost of energy increases to a sufficiently high value.

当能源成本很高时，在传统操作中更多地使用这类电机必然变得很重要。

3. The only difference is in the control and shape of the excitation voltage: an AC synchronous motor is fed with more or less sinusoidal waveforms which in turn produce a rotating magnetic field.

唯一的区别是励磁电压的控制和形状：交流同步电动机采用正弦波或近似正弦波供电，这些波形进而产生旋转磁场。

Exercises

1. Answer the following questions according to the text

(1) The excitation end does not consume electric energy for permanent magnet motors. True or false?

(2) What is in common about the structure design of PM DC brushless and AC synchronous motor?

(3) What is difference between PMDC brushless and AC synchronous motor?

(4) Why is the power density of the brushless motor higher than that of DC commutator motor?

(5) What kind of motoris considered as the best driving motor for electric and hybrid vehicles?

2. Translate the following sentences into Chinese according to the text

(1) One such application is in servo drives in which the motor is commanded to operate for a long period at, or very near, zero speed.

(2) The construction of a PM DC commutator motor is similar to a DC motor with the electromagnetic excitation system replaced by PMs.

(3) The armature current of synchronous and DC brushless motors is not transmitted through brushes, which are subject to wear and require maintenance.

(4) In addition, considerable improvements in dynamics can be achieved because the air

gap magnetic flux density is high, the rotor has a lower inertia and there are no speed-dependent current limitations.

(5) The number of auxiliary DC PM commutator motors can vary from a few in an inexpensive car to about one hundred in a luxury car.

3. Translate the following paragraph into Chinese

Rare earth PMs can not only improve the motor's steady state performance but also the power density (output power-to-mass ratio), dynamic performance, and quality. The prices of rare earth magnets are also dropping, which is making these motors more popular. The improvements made in the field of semiconductor drives have meant that the control of brushless motors has become easier and cost effective, with the possibility of operating the motor over a large range of speeds while still maintaining good efficiency.

Text B Permanent Magnet Synchronous Motors (PMSM)

1. Construction

Recent developments in rareearth PM materials and power electronics have opened new prospects on the design, construction and application of PM synchronous motors (PMSMs). Servo drives with PM motors fed from static inverters are finding applications on an increasing scale. PM servo motors with continuous output power of up to 15 kW at 1500 r/min are common. Commercially, PM AC motor drives are available with ratings up to at least 746 kW. Rare earth PMs have also been recently used in large power synchronous motors rated at more than 1 MW. Large PM motors can be used both in low-speed drives (ship propulsion) and high-speed drives (pumps and compressors).

PM synchronous motors are usually built with one of the following rotor configurations:

(a) Classical, with salient poles, laminated pole shoes and a cage winding Fig. 1. 9 (a).

(b) Interior-magnet rotor Fig. 1. 9 (b).

(c) Surface-magnet rotor Fig. 1. 9 (c).

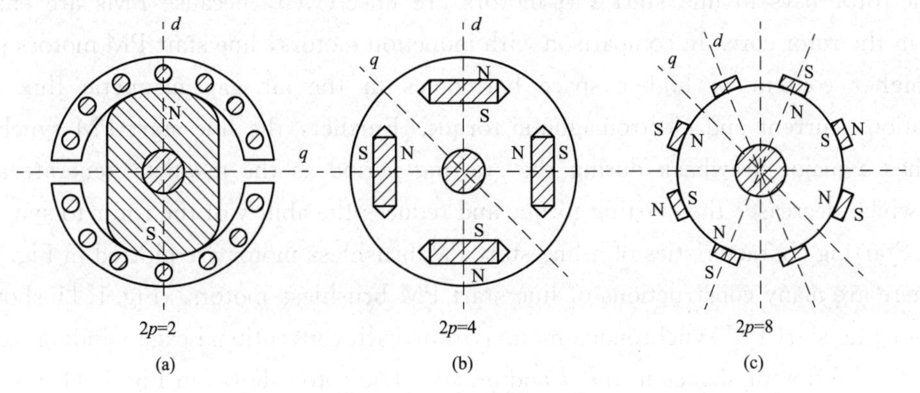

Fig. 1. 9 Rotor configurations for PM synchronous motors (—)

(a) classical configuration; (b) interior-magnet rotor; (c) surface-magnet rotor;

(d) inset-magnet rotor Fig. 1. 9 (d).

(e) rotor with buried magnets symmetrically distributed Fig. 1. 9 (e).

(f) rotor with buried magnets asymmetrically distributed Fig. 1. 9 (f).

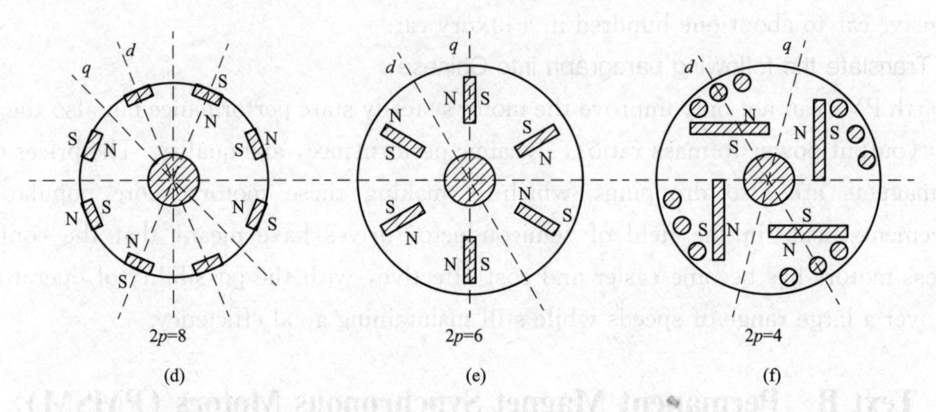

Fig. 1. 9 Rotor configurations for PM synchronous motors (二)

(d) inset-magnet rotor; (e) rotor with buried (spoke) magnets symmetrically distributed;

(f) rotor with buried magnets asymmetrically distributed.

2. Starting

(1) Asynchronous starting

A synchronous motor is not self-starting. To produce an asynchronous starting torque, its rotor must be furnished with a cage winding or mild steel pole shoes. The starting torque is produced as a result of the interaction between the stator rotating magnetic field and the rotor currents induced in the cage winding or mild steel pole shoes.

PM synchronous motors that can produce asynchronous starting torque are commonly called line start PM synchronous motors (LS-PMSMs). These motors can operate without solid state converters. After starting, the rotor is pulled into synchronism and rotates with the speed imposed by the line input frequency. The efficiency of line start PM motors is higher than that of equivalent induction motors and the power factor can be equal to unity.

The rotor bars in line start PM motors are unscrewed, because PMs are embedded axially in the rotor core. In comparison with induction motors, line start PM motors produce much higher content of higher space harmonics in the air gap magnetic flux density distribution, current and electromagnetic torque. Further, the line start PM synchronous motor has a major drawback during the starting period as the magnets generate a brake torque which decreases the starting torque and reduces the ability of the rotor to synchronize a load. Starting characteristics of a line start PM brushless motor are plotted in Fig. 1. 10.

There are many constructions of line start PM brushless motors. Fig. 1. 11 shows two rotors for line start PM synchronous motor: rotor with conventional cage winding and rotor with slots of different shapes in the d and q-axis. The rotor shown in Fig. 1. 11 (b) allows for significant reduction of the 5th, 11th, 13th, 17th and higher odd harmonics.

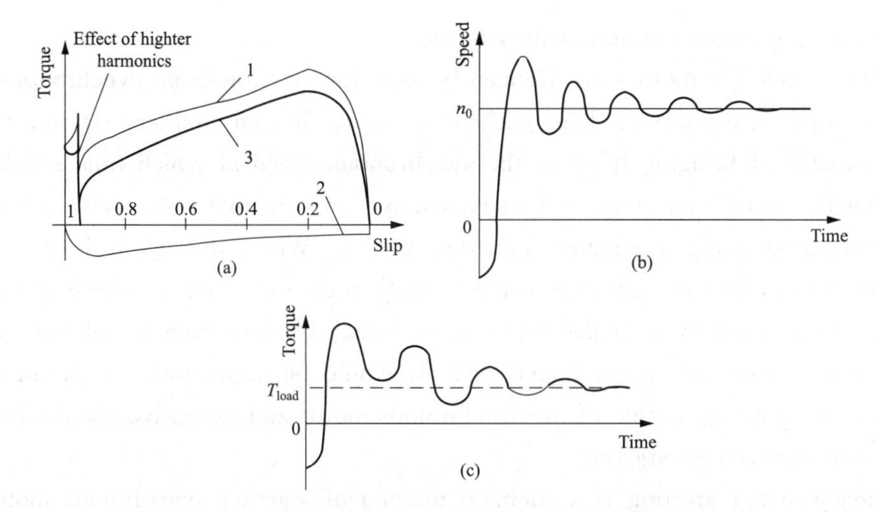

Fig. 1. 10 Characteristics of a line start PM brushless motor

(a) steady-state torque-slip characteristic; (b) speed-time characteristic; (c) torque-time characteristic

1—asynchronous torque; 2—braking torque produced by PMs; 3—resultant torque, n_0—steady-state speed, T_{load}—load torque

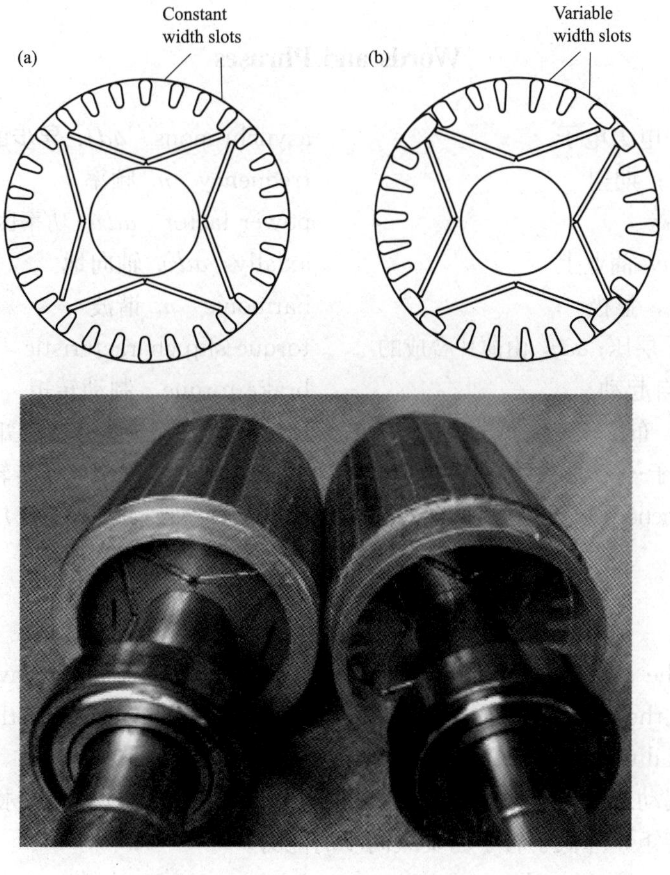

Fig. 1. 11 Rotors of line start PM synchronous motors

(a) constant width slots; (b) variable width slots

(2) Starting by means of an auxiliary motor

Auxiliary induction motors are frequently used for starting large synchronous motors with electromagnetic excitation. The synchronous motor has an auxiliary starting motor on its shaft, capable of bringing it up to the synchronous speed at which time synchronizing with the power circuit is possible. The unexcited synchronous motor is accelerated to almost synchronous speed using a smaller induction motor. When the speed is close to the synchronous speed, first the armature voltage and then the excitation voltage is switched on, and the synchronous motor is pulled into synchronism. The disadvantage of this method is it's impossible to start the motor under load. It would be impractical to use an auxiliary motor of the same rating as that of the synchronous motor and expensive installation.

(3) Frequency-change starting

Frequency-change starting is a common method of starting synchronous motors with electromagnetic excitation. The frequency of the voltage applied to the motor is smoothly changed from the value close to zero to the rated value. The motor runs synchronously during the entire starting period being fed from a variable voltage variable frequency (VVVF) solid state inverter.

Words and Phrases

power electronic 电力电子
prospect *n.* 前景；期望
inverter *n.* 逆变器
commercially *adv.* 商业上
salient pole 凸极，显极
laminate *n.* 叠层，层压；*adj.* 由薄片叠成的
self-starting *n.* 自起动
furnish *v.* 陈设，布置；提供，供应
as a result of 由于……的结果
stator rotating magnetic field 定子旋转磁场
asynchronous *adj.* 异步的
frequency *n.* 频率
power factor *adv.* 功率因数
axially *adv.* 轴向地
harmonic *n.* 谐波
torque-slip characteristic 转矩-转差率特性
brake torque 制动转矩
resultant torque 合成转矩
synchronous speed 同步转速
accelerate *v.* 加快，(使)增速；加速

Notes

1. Further, the line start PM synchronous motor has a major drawback during the starting period as the magnets generate a brake torque which decreases the starting torque and reduces the ability of the rotor to synchronize a load.

此外，异步起动永磁同步电动机在起动过程中有一个主要的缺点：永磁体产生制动转矩，减小了起动转矩，降低了转子与负载同步的能力。

2. The synchronous motor has an auxiliary starting motor on its shaft, capable of bringing it up to the synchronous speed at which time synchronizing with the power circuit is possible.

同步电动机在其轴上装有辅助起动电机，使带着它达到与电网一致的同步转速成为可能。

Exercises

1. Answer the following questions according to the text

(1) What promotes the development of permanent magnet synchronous motors (PMSMs) in recent years?

(2) What is the maximum power factor of LS-PMSMs?

(3) What are the characteristics in air gap magnetic field of LS-PMSMs when compared with asynchronous motor?

(4) A large synchronous motor without PM excitation is able to be started with load by means of an auxiliary motor. True or false?

(5) Some measures must be executed in the rotor to start a permanent magnet synchronous motor (PMSM). True or false?

2. Translate the following sentences into Chinese according to the text

(1) The starting torque is produced as a result of the interaction between the stator rotating magnetic field and the rotor currents induced in the cage winding or mild steel pole shoes.

(2) After starting, the rotor is pulled into synchronism and rotates with the speed imposed by the line input frequency.

(3) When the speed is close to the synchronous speed, first the armature voltage and then the excitation voltage is switched on, and the synchronous motor is pulled into synchronism.

(4) It would be impractical to use an auxiliary motor of the same rating as that of the synchronous motor and expensive installation.

(5) The motor runs synchronously during the entire starting period being fed from a variable voltage variable frequency (VVVF) solid state inverter.

3. Translate the following paragraph into Chinese

An important consequence of the method of mounting the rotor magnets is the difference between direct and quadrature axes inductance values. The rotor magnetic axis is called direct axis and the principal path of the flux is through the magnets. Consider the permeability of high flux density PMs is almost that of the air. This results in the magnet thickness becoming an extension of air gap by that amount. The stator inductance when the direct axis or magnets are aligned with the stator winding is known as direct axis inductance. By rotating the magnets from the aligned position by 90°, the stator flux sees the interpolar area of the rotor containing only the iron path and the inductance measured in this position is referred to as quadrature axis inductance.

Unit 5 Electrical Apparatus in Control System

Text A Relays

Relays are extremely useful when we have a need to control a large amount of current and/or voltage with a small electrical signal. The relay coil which produces the magnetic field may only consume fractions of a watt of power, while the contacts closed or opened by that magnetic field may be able to conduct hundreds of times that amount of power to a load. In effect, a relay acts as a binary (on or off) amplifier.

In Fig. 1. 12, the relay's coil is energized by the low-voltage (12V DC) source, while the single-pole, single-throw (SPST) contact interrupts the high-voltage (380V AC) circuit. It is quite likely that the current required to energize the relay coil will be hundreds of times less than the current rating of the contact. Typical relay coil currents are well below 1 A (amp), while typical contact ratings for industrial relays are at least 10 A.

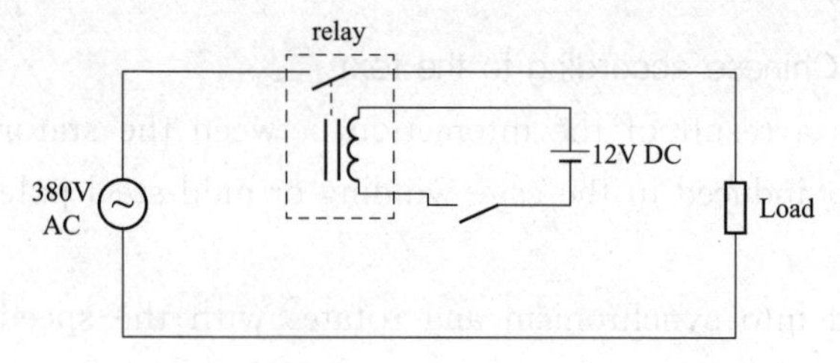

Fig. 1. 12 Schematic circuit diagram of relay

One relay coil/armature assembly may be used to actuate more than one set of contacts. Those contacts may be normally-open, normally-closed, or any combination of the two. As with switches, the "normal" state of a relay's contacts is that state when the coil is de-energized, just as you would find the relay sitting on a shelf, not connected to any circuit.

The choice of contacts in a relay depends on the same factors which dictate contact choice in other types of switches. Open-air contacts are the best for high-current applications, but their tendency to corrode and spark may cause problems in some industrial environments. Mercury and reed contacts are sparkless and won't corrode, but they tend to be limited in current-carrying capacity.

Fig. 1. 13 are three small relays (about two inches in height, each), installed on a panel as part of an electrical control system at a municipal water treatment plant:

The relay units shown here are called "octal-base", because they plug into matching sockets, the electrical connections secured via eight metal pins on the relay bottom. The screw terminal connections you see in the photograph where wires connect to the relays are actually part of

Fig. 1. 13 Photo of three small relays

the socket assembly, into which each relay is plugged. This type of construction facilitates easy removal and replacement of the relay (s) in the event of failure.

Aside from the ability to allow a relatively small electric signal to switch a relatively large electric signal, relays also offer electrical isolation between coil and contact circuits. This means that the coil circuit and contact circuit (s) are electrically insulated from one another. One circuit may be DC and the other AC, and/or they may be at completely different voltage levels, across the connections or from connections to ground.

While relays are essentially binary devices, either being completely on or completely off, there are operating conditions where their state may be indeterminate, just as semiconductor logic gates. In order for a relay to positively "pull in" the armature to actuate the contact (s), there must be a certain minimum amount of current through the coil. This minimum amount is called the pull-in current, and it is analogous to the minimum input voltage that a logic gate requires to guarantee a "high" state (typically 2V Volts for TTL, 3.5V Volts for CMOS). The coil current must drop below a value significantly lower than the pull-in current before the armature "drops out" to its spring-loaded position and the contacts resume their normal state. This current level is called the drop-out current, and it is analogous to the maximum input voltage that a logic gate input will allow to guarantee a "low" state (typically 0.8 V for TTL, 1.5 V for CMOS).

The hysteresis, or difference between pull-in and drop-out currents, results in operation that is similar to a Schmitt trigger logic gate. Pull-in and drop-out currents (and voltages) vary widely from relay to relay, and are specified by the manufacturer.

Words and Phrases

electrical apparatus 电器
relay *n.* 继电器
electrical signal 电信号
consume *v.* 消耗
contact *n.* 触点，触头
binary *adj.* 二进制的
amplifier *n.* 放大器
single-pole *adj.* 单刀
single-throw *adj.* 单掷
amp. 安培（同 ampere）
normally-open *adj.* 常开
normally-closed *adj.* 常闭
corrode *v.* 使腐蚀
spark *v.* 发出火花；*n.* 火花
sparkless *adj.* 无火花的
current-carrying capacity 载流能力
mercury *n.* 水银
reed *n.* 簧片
panel *n.* 面板
municipal *adj.* 市立的，地方性的
water treatment plant 水处理车间
socket *n.* 插座
pin *n.* 插脚
screw *n.* 螺钉
terminal connections 接线端
semiconductor *n.* 半导体
logic gate 逻辑门
resume *v.* 恢复
hysteresis *n.* 滞回现象
armature *n.* （电器的）衔铁
pull-in current 动作电流
drop-out current 返回电流

Schmitt trigger　施密特触发器

Notes

1. Aside from the ability to allow a relatively small electric signal to switch a relatively large electric signal, relays also offer electrical isolation between coil and contact circuits. This means that the coil circuit and contact circuit (s) are electrically insulated from one another.

继电器除具有用一个小电信号开关一个大电信号的能力以外，还可以在线圈电路与触点电路之间起电气隔离的作用，这说明驱动线圈电路与触点电路之间是互相隔离的。

2. In order for a relay to positively "pull in" the armature to actuate the contact (s), there must be a certain minimum amount of current through the coil. This minimum amount is called the pull-in current, and it is analogous to the minimum input voltage that a logic gate requires to guarantee a "high" state (typically 2 V for TTL, 3.5 V for CMOS).

为了保证使继电器的线圈能够驱动触点而在线圈中流过的最小电流称为动作电流。这与为保证逻辑门有高电平而需施加的最小电压（TTL 的典型值是 2 V，CMOS 为 3.5 V）相似。

Exercises

1. Answer the following questions according to the text

(1) Relay acts as an octal amplifier. Right or wrong?

(2) The "normal" state of a relay's contacts is the state when the coil is de-energized. Right or wrong?

(3) The coil circuit and contact circuit (s) of relay, one circuit is DC and the other is AC, is this case allowable?

(4) What is analogous to relay and logic gate?

(5) The pull-in current and the pull-out current of a relay may be the same value. Right or wrong?

(6) The pull-in current may be different for different relays. Right or wrong?

2. Translate the following sentences into Chinese according to the text

(1) The relay coil which produces the magnetic field may only consume fractions of a watt of power, while the contacts closed or opened by that magnetic field may be able to conduct hundreds of times that amount of power to a load.

(2) The contacts of a relay may be normally-open, normally-closed, or any combination of the two.

(3) This type of construction facilitates easy removal and replacement of the relay (s) in the event of failure.

(4) While relays are essentially binary devices, either being completely on or completely off, there are operating conditions where their state may be indeterminate, just as with

semiconductor logic gates.

(5) Pull-in and drop-out currents (and voltages) vary widely from relay to relay, and are specified by the manufacturer.

3. Translate the following paragraph into Chinese

Relay is a key component in any power system protection scheme. It is a device that, based on information received from the power system, performs one or more switching actions. The information referred to consists of signals and currents, typically the output of instrument transformers (互感器). The relay decides to close (or open) one or more sets of normally open (or close) contacts. The switching action typically energizes the trip coil of a circuit breaker, which then opens the power circuit.

Text B Contactors

Contactors typically have multiple contacts, and those contacts are usually (but not always) normally-open, so that power to the load is shut off when the coil is de-energized. Perhaps the most common industrial use for contactors is the control of electric motors.

In Fig. 1. 14, the top three contacts switch the respective phases of the incoming 3-phase AC power, typically at least 380 V for motors. The lowest contact is an "auxiliary" contact which has a current rating much lower than that of the large motor power contacts, but is actuated by the same armature as the power contacts. The auxiliary contact is often used in a relay logic circuit, or for some other part of the motor control scheme. One contactor may have several auxiliary contacts, either normally-open or normally-closed, if required.

The three " opposed-question-mark " shaped devices in series with each phase going to the motor are called overload heaters. If the temperature of any of these heater elements reaches a critical point (equivalent to a moderate overloading of the motor), a normally-closed switch contact (not shown in the diagram) will spring open. This normally-closed contact is usually connected in series with the contactor coil, so that when it opens the contactor will automatically de-energize, thereby shutting off power to the motor. Overload heaters are intended to provide overcurrent protection for large electric motors, unlike circuit breakers and fuses which serve the primary purpose of providing overcurrent protection for power devices.

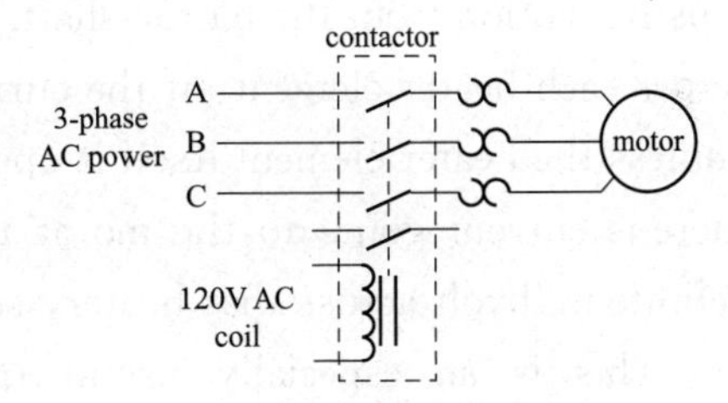

Fig. 1. 14 Schematic circuit diagram of contactor

Fig. 1. 15 is a contactor for a three-phase electric motor, installed on a panel as part of an electrical control system at a municipal water treatment plant. Three-phase, 380 V AC power comes in to the three normally-open contacts at the top of the contactor via screw terminals labeled "L1" "L2" and "L3". Power to the motor exits the overload heater assembly at the bottom of this device via screw terminals labeled "T1" "T2" and "T3."

Fig. 1. 15 Photo of contactor for a three-phase electric motor

The overload heater units themselves are black, square-shaped blocks with the label "W34" indicating a particular thermal response for a certain horsepower and temperature rating of electric motor. If an electric motor of differing power and/or temperature ratings were to be substituted for the one presently in service, the overload heater units would have to be replaced with units having a thermal response suitable for the new motor. The motor manufacturer can provide information on the appropriate heater units to use.

A white pushbutton located between the "T1" and "T2" line heaters serves as a way to manually re-set the normally-closed switch contact back to its normal state after having been tripped by excessive heater temperature. Wire connections to the "overload" switch contact may be seen at the lower-right of the photograph, near a label reading "NC" (normally-closed). On this particular overload unit, a small "window" with the label "Tripped" indicates a tripped condition by means of a colored flag. In this photograph, there is no "tripped" condition, and the indicator appears clear.

How do you know if the motor is consuming power when the contactor coil is energized and the armature has been pulled in? If the motor's windings are burnt open, you could be sending voltage to the motor through the contactor contacts, but still have zero current, and thus no motion from the motor shaft. You can take your multimeter and measure millivolt across each heater element: if the current is zero, the voltage across the heater will be zero (unless the heater element itself is open, in which case the voltage across it will be large); if there is current going to the motor through that phase of the contactor, you will read a definite millivolt across that heater, see Fig. 1. 16.

This is an especially useful trick to use for troubleshooting 3-phase AC motors, to see if one phase winding is burnt open or disconnected, which will result in a rapidly destructive condition. If one of the lines carrying power to the motor is open, it will not have any current through it (as indicated by a 0. 00 mV reading across its heater), although the other two lines will (as indicated by small amounts of voltage dropped across the respective heaters).

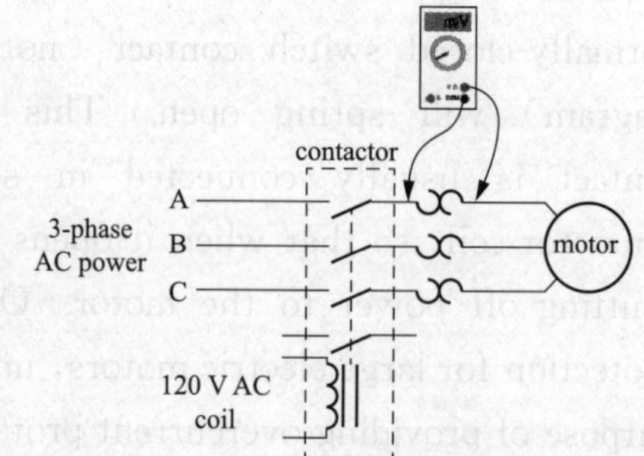

Fig. 1. 16 Measure the millivolt across the heater element

A contactor is a large relay, usually used to switch current to an electric motor or other high-power load. Large electric motors can be protected from overcurrent damage through the use of overload heaters and overload contacts. If the series-connected heaters get too hot from excessive current, the normally-closed overload contact will open, de-energizing the contactor sending power to the motor.

Words and Phrases

contactor *n.* 接触器
multiple *n.* 若干
auxiliary contact 辅助触点
current rating 电流等级
de-energize *v.* 使失电，使失磁
in series with 与……串联
overload *n.* 过负荷，过载
heater *n.* 发热器，加热器
shunt resistor 分路电阻器，并联电阻器
pushbutton *n.* 按钮
trip *v.* 脱扣
lower-right *n.* 右下方
multimeter *n.* 万用表
millivolt *n.* 毫伏
trick *n.* 窍门
troubleshooting *n.* 发现故障并维修
burnt open 断路，断开

Notes

1. The top three contacts switch the respective phases of the incoming 3-phase AC power, typically at least 380 V for motors. The lowest contact is an "auxiliary" contact which has a current rating much lower than that of the large motor power contacts, but is actuated by the same armature as the power contacts.

上面的三个触点分别开关三相交流电源，对三相电动机此电压的典型值为380V。最下面的触点是辅助触点，其电流额定值远低于大电动机主电路触点的电流额定值，主触点和辅助触点由同一电枢绕组激励。

2. This is an especially useful trick to use for troubleshooting 3-phase AC motors, to see if one phase winding is burnt open or disconnected, which will result in a rapidly destructive condition.

这是发现并解决三相交流电动机故障的一个非常有用的窍门，看是否有一相绕组断路造成迅速损坏的状况。

Exercises

1. Answer the following questions according to the text

(1) How many auxiliary contacts may a contactor have?

(2) Where are contactors often used in?

(3) How do you know if the motor is consuming power when the contactor coil is energized?

(4) A contactor is a large relay. Right or wrong?

2. Translate the following sentences into Chinese according to the text

(1) Contactors typically have multiple contacts, and those contacts are usually (but not always) normally-open, so that power to the load is shut off when the coil is de-energized.

(2) If the temperature of any of these heater elements reaches a critical point (equivalent to a moderate overloading of the motor), a normally-closed switch contact (not shown in the diagram) will spring open.

(3) A white pushbutton located between the "T1" and "T2" line heaters serves as a way to manually re-set the normally-closed switch contact back to its normal state after having been tripped by excessive heater temperature.

(4) A contactor is a large relay, usually used to switch current to an electric motor or other high-power load.

3. Translate the following paragraph into Chinese

A contactor is a large relay, usually used to switch current to an electric motor or other high-power load. Large electric motors can be protected from overcurrent damage through the use of overload heaters and overload contacts. If the series-connected heaters get too hot from excessive current, the normally-closed overload contact will open, de-energizing the contactor sending power to the motor.

Unit 6 Electrical Apparatus Reliability

Test A Electrical Apparatus Reliability Tests

Electrical apparatus reliability is referred to the apparatus' ability to achieve the prescribed functions within prescribed time under prescribed conditions. It reflects an apparatus' good performances in failure-free, maintainability, durability, validity, economy-utility and longevity. Thus, reliability is an important quality index.

Apparatus reliability data can be achieved through tests in lab or field survey. In recent 20 years, such reliability test researches in China are almost confined in labs. The products that have been carried out such tests involve electromagnetism auxiliary relay, small load AC contactor, fuse, circuit-breaker, electric leakage switch, and so on. Such researches are effective to improve low voltage apparatus quality. Carrying such reliability tests in lab has some advantages: such as the same test conditions for test product, the same failure judgment standards, which makes it easier to compare the test results, and determine the proper reliability index and judging method. But the testing conditions are not the same with product using conditions. The reliability tests of the main technique performances for some products cost too much. And for the limitations of expenses and the number of some valuable products, it is almost impossible to do complete reliability tests. For example, the reliability tests for contactor only test the main circuits and non-load operating performance of some small load AC contactor. And the tests of longevity and make-break capability cannot be carried out due to the high expenses. The reliability tests of fuse only can test the current-time influences on aging of fuse caused by the starting current of electromotor. And the tests on its breaking capability cannot be carried out. The reliability tests of breakers only can test the operations of non-load release. And as for electricity protective characteristics, longevity and utmost breaking capability are not involved.

Electrical apparatus operate under different operating environments. The reliability test in lab cannot consider all possible conditions including temperature, humidity, impact, vibration, harmful media etc. Whereas for field survey, through choosing the location of field survey, the reliability under all kinds of practical conditions can be reflected clearly. The reliability design of a set of low-voltage electrical apparatus needs complete reliable data of all components, which would be limited to achieve in lab. But through field survey, the tests can reflect the practical operating circumstances and provide reliable data for design.

The traditional apparatus reliability tests carried out in lab have some limitations, such as too strict test conditions or so less outputs. For the valuable products, to carry reliability tests at the operating location is much better than in the lab. To ensure the high reliability during the use of such apparatus and to avoid losses taken by maintaining or changing parts after failure appears, collecting and analyzing the data of the apparatus reliability indexes during its normal use become very important. Mastering the invalidate data and then giving precaution before apparatus fails are necessary.

Biology Immune System is a complex system composed by organs, cells, and molecules. Except for nerve system it's another one that the organism can recognize the stimulation from "itself" and "non-itself", response precisely and reserve the memory. It has the characteristics such as pattern recognition, self-learning, memory achieving, distributed detecting and so on. Artificial immune system, which is developed from biology immune system, can recognize and process the input information, learn the right input, produce immunity and respond to the abnormal input. Taking the reliability data collected from the spot as the input for artificial immune system, the system can judge the abnormal operating state of apparatus. These make it possible to maintain or change the parts of electrical apparatus in time and avoid losses. So, artificial immune system is a novel method of electrical apparatus reliability test.

Words and Phrases

electrical apparatus 电器
reliability *n.* 可靠性
failure-free *n.* 无故障性
maintainability *n.* 维修性
durability *n.* 耐久性
validity *n.* 有效性
economy-utility *n.* 使用经济性
index *n.* 指标
fuse *n.* 熔断器
field survey 现场调查
electromagnetism auxiliary relay 电磁式中间继电器
small load AC contactor 小容量交流接触器
electric leakage switch 漏电开关
circuit-breaker *n.* 断路器
electromotor *n.* 电动机
longevity *n.* 寿命
make-break capability 分断能力
humidity *n.* 湿度
impact *n.* 冲击，撞击
vibration *n.* 振动
harmful media 有害介质
aging *n.* 老化
precaution *n.* 预警，防范
spot *n.* 现场
biology immune 生物免疫
nerve system 神经系统
characteristic *n.* 特性，特征
pattern recognition 模式识别
self-learning *n.* 自学习
distributed detecting 分布式检测
abnormal *adj.* 反常的
novel *adj.* 新颖的

Notes

1. Electrical apparatus reliability is referred to the apparatus ability to achieve the prescribed functions within prescribed time under prescribed conditions.

电器的可靠性是指电器产品在规定的条件下和规定的时间内完成规定功能的能力。

2. Carrying such reliability tests in lab has some advantages: such as the same test conditions for test product, the same failure judgment standards, which make it easier to compare the test results, determine the proper reliability index and judge method.

在实验室内进行可靠性试验研究，具有试品的试验条件和故障评判标准统一、可比性强以及便于确定可靠性考核指标和评定方法的优点。

3. Except for nerve system it's another one that the organism can recognize the stimulation from "itself" and "non-itself", response precisely and reserve the memory.

它是除神经系统外，机体能特异地识别“自己”和“非己”刺激，对之做出精确应答并保留记忆的功能系统。

Exercises

1. Answer the following questions according to the text

(1) What is electrical apparatus reliability?

(2) Apparatus reliability data can only be achieved through tests in lab. Right or wrong?

(3) Where the traditional apparatus reliability tests carried out in?

(4) Biology Immune System has the characteristics such as pattern recognition, self-learning, memory achieving, distributed detecting. Right or wrong?

(5) Artificial immune system is one of the methods of electrical apparatus reliability test. Right or wrong?

2. Translate the following sentences into Chinese according to the text

(1) The products that have been carried out such tests involve electromagnetism auxiliary relay, small load AC contactor, fuse, circuit-breaker, electric leakage switch, and so on.

(2) The reliability tests of the main technique performances for some products costs too much.

(3) And for the limitations of expenses and the number of some valuable products, it is almost impossible to do complete reliability tests.

(4) The reliability test in lab cannot consider all possible conditions including temperature, humidity, impact, vibration, harmful media etc.

(5) Mastering the invalidate data and then giving precaution before apparatus fails are necessary.

3. Translate the following paragraph into Chinese

In order to confirm and guarantee the reliability of electrical contacts, a lot of experience is required, especially on how to develop additional testing methods on top of the requirements given though national and international standards. It is very important that the manufacturers of devices with electrical contacts are developing such methods (temperature rise tests, short circuit tests, high voltage tests, etc.) and continuously are using them as routine tests in order to confirm the reliability of electrical contacts.

Text B Influencing Factor Analysis of Circuit-Breaker Reliability

The aim of the research on electrical apparatus reliability is to improve the product quality, increase the economic benefit, raise the product technique level, heighten the scientific research level, finally reduce or prevent the loss and injury caused by the product failure. So far as the reliability of the circuit-breaker is concerned, it is the best to prolong the accidental failure period, i. e. prolong the service life within the permitted expenses as far as possible. For this reason, first to get rid of the early period failure, this is what the product manufacturer should and must perform; secondly, before the failure rate goes up, the parts and components which are easy to break down should be changed. These can prolong the service life of the product, or changing and installing the new circuit-breaker, so as to assure the safety of the distribution network and the person.

According to the above-mentioned reliability theory analysis, as viewed from the circuit-breaker structural principle, the factors of influencing the circuit-breaker early period reliability are summarized as the following.

(1) Influence of Arc-Suppressing Performance: The arc-suppression of the circuit-breaker relies mainly on the grid plate in the arc-suppressing chamber to divide the electric arc, speedily cooling and extinguishing at zero-crossing. In view of the design stereotype of the arc-suppressing chamber, its arcing distance, arc-extinguishing time, high-temperature resistance and insulating property will all influence the reliability of the circuit-breaker, and these performances will mainly depend on the materials and the assembly quality. In order to enhance the quality of the product, the improvement of the design should also be performed for the arc-suppressing chamber, such as the arc-suppressing chamber without arcing.

(2) Influence of Contact System: The contact of the circuit-breaker is not only to have the function of conducting electricity but also open and close the electric circuit. And the latter is more important. The contact consists of the various parts of the silver alloying contact based on the copper, the weaving copper line and the soft connection etc., when assembling into the complete appliance, also including the integral components of the support frame, the spring and the shaft etc. Therefore, the contact material, dimensional shape and its welding (or riveting) quality will have the larger influence on the reliability of the circuit-breaker. The phenomena are shown as the following: high temperature rise, poor anti-fusion welding and arc-resistance capability, tripping capability being not up to standard requirements etc.

(3) Influence of Mechanism System: The mechanism of the circuit-breaker is the energy storage type for connecting rod mechanisms, what determined its reliability mainly is the anti-fatigue strength of the spring, the rigidity of the shaft and the connecting rod etc.

(4) Influence of Tripping System: The tripping system of the circuit-breaker, regardless of the heat tripping (overload protection) or the electromagnetic tripping (instantaneous protection),

firstly depends on the materials and the assembly quality. For example, the heat double metal plate's specific curvature and the heat stability will directly influence the reliability of the overload protection of the circuit-breaker (including accuracy and consistency). Secondly, there also are the adjustment-test and the inspection before the product leaving the factory. Whether or not the tripping system of the circuit-breaker can complete the stipulated protective function, and not influencing the normal usage and production, these will depend on the adjustment-test's setting value and the delay feature. If the productive adjustment-test method, parameters or the equipment performance cannot meet the requirements, there is no way to discuss the accurate and reliable protective feature.

(5) Influence of Case: The case of the circuit-breaker includes the cover, the basement and the flash barrier etc., mainly playing the role of supporting other parts, insulation and seal. The case material quality and the case machining quality determine its mechanical performance and insulation property, thus influencing the reliability of the circuit-breaker.

(6) Influence of Other Factors: The selection of the circuit-breaker, installation, usage, maintenance and different working systems will influence its reliability.

Words and Phrases

influencing factor 影响因素
product quality 产品质量
failure *n.* 故障
early period failure 早期失效
distribution network 配电网络
service life 使用寿命
arc-suppression *n.* 灭弧
arc-extinguishing *n.* 灭弧
welding *n.* 焊接
riveting *n.* 铆接
grid plate 栅片
arc-suppressing chamber 灭弧室
extinguishing at zero-crossing 过零熄火
anti-fusion welding 抗熔焊性
arc-resistance capability 耐弧能力
connecting rod 连杆
anti-fatigue strength 抗疲劳强度
spring *n.* 弹簧
trip *v.* 脱扣
assembly quality 装配质量
double metal plate's specific curvature 双金属片的比弯曲
adjustment-test *n.* 调试
inspection *n.* 检验
stipulated protective function 规定的保护功能
case *n.* 外壳
cover *n.* 盖子
basement *n.* 基座
flash barrier *n.* 隔弧板
seal *n.* 密封
maintenance *n.* 维护
setting value 整定值

Notes

1. So far as the reliability of the circuit-breaker is concerned, it is the best to prolong the

accidental failure period, i. e. prolong the service life within the permitted expenses as far as possible.

从断路器的可靠性来说，最好延长偶然失效期，即尽量在允许的费用内延长使用寿命。

2. In view of the design stereotype of the arc-suppressing chamber, its arcing distance, arc-extinguishing time, high-temperature resistance and insulating property will all influence the reliability of the circuit-breaker, and these performances will mainly depend on the materials and the assembly quality.

由于灭弧室本身设计定型，其飞弧距离、熄弧时间、耐高温、绝缘性能等都会影响到断路器的可靠性，而这些性能主要取决于材料和转配质量。

3. The contact consists of the various parts of the silver alloying contact based on the copper, the weaving copper line and the soft connection etc., when assembling into the complete appliance, also including the integral components of the support frame, the spring and the shaft etc.

触头由铜基、银金属触头、铜编织线软连接等部分组成，装入整机时还有支架、弹簧、轴等配套零部件。

Exercises

1. Answer the following questions according to the text

(1) what should the product manufacturer perform?

(2) In view of the design stereotype of the arc-suppressing chamber, which factors will influence the reliability of the circuit-breaker?

(3) What are the functions of the contact of circuit-breaker?

(4) What will do before the circuit-breaker leaving the factory?

(5) The selection of the circuit-breaker, installation, usage, maintenance and different working systems will influence its reliability. Right or wrong?

2. Translate the following sentences into Chinese according to the text

(1) Before the failure rate goes up, the parts and components which are easy to break down should be changed.

(2) These can prolong the service life of the product, or changing and installing the new circuit-breaker, so as to assure the safety of the distribution network and the person.

(3) The contact material, dimensional shape and its welding (or riveting) quality will have the larger influence on the reliability of the circuit-breaker.

(4) The heat double metal plate's specific curvature and the heat stability will directly influence the reliability of the overload protection of the circuit-breaker.

(5) The case material quality and the case machining quality determine its mechanical performance and insulation property, thus influencing the reliability of the circuit-breaker.

3. Translate the following paragraph into Chinese

The reliability means the ability that the product fulfills the stipulated conditions. The reliability working scope is quite large, for example: the reliability organization management, the reliability data, the reliability inspection and the reliability theory etc. The factors of influencing the circuit-breaker reliability can give a reference to improve the circuit-breaker quality.

Part Ⅱ Theory and New Technology of Electrical Engineering

Unit 1 Magnetism and Electromagnetism

Text A Electromagnetism

The discovery of the relationship between magnetism and electricity, like many other scientific discoveries, was stumbled upon almost by accident. The Danish physicist Hans Christian Oersted was lecturing one day in 1820 on the possibility of electricity and magnetism being related to one another, and in the process demonstrated it conclusively by experiment in front of his whole class! By passing an electric current through a metal wire suspended above a magnetic compass, Oersted was able to produce a definite motion of the compass needle in response to the current. What began as conjecture at the start of the class session was confirmed as a fact at the end. His serendipitous discovery paved the way for a whole new branch of science: electromagnetics.

Detailed experiments showed that the magnetic field produced by an electric current is always oriented perpendicular to the direction of flow. A simple method of showing this relationship is called the right-hand rule. Simply stated, the right-hand rule says that the magnetic flux lines produced by a current-carrying wire will be oriented the same direction as the curled fingers of a person's right hand, with the thumb pointing in the direction of electron flow, which is shown in Fig. 2. 1.

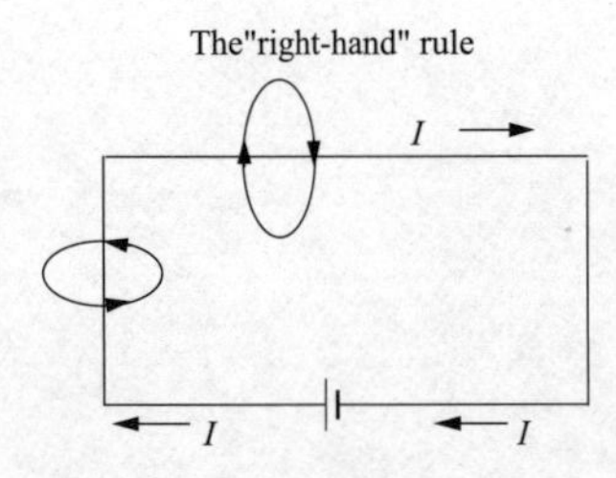

Fig. 2. 1 Magnetic field produced by an electric current

The magnetic field encircles this straight piece of current-carrying wire, the magnetic flux lines having no definite "north" or "south" poles.

While the magnetic field surrounding a current-carrying wire is indeed interesting, it is quite weak for common amounts of current, able to deflect a compass needle and not much more. To create a stronger magnetic field force with the same amount of electric current, we can wrap the wire into a coil shape, where the circling magnetic fields around the wire will join to create a larger field with a definite magnetic (north and south) polarity. See Fig. 2. 2.

The amount of magnetic field force generated by a coiled wire is proportional to the current through the wire multiplied by the number of "turns" or "wraps" of wire in the coil. This field force is called magnetomotive force (MMF), and is very much analogous to electromotive force (EMF) in an electric circuit.

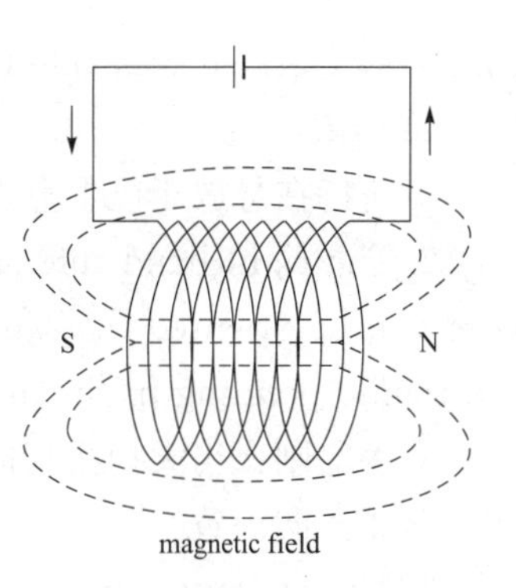

Fig. 2. 2 Circling magnetic fields around the wire

An electromagnet is a piece of wire intended to generate a magnetic field with the passage of electric current through it. Though all current-carrying conductors produce magnetic fields, an electromagnet is usually constructed in such a way as to maximize the strength of the magnetic field it produces for a special purpose. Electromagnets find frequent application in research, industry, medical, and consumer products.

As an electrically-controllable magnet, electromagnets find application in a wide variety of "electromechanical" devices: machines that effect mechanical force or motion through electrical power. Perhaps the most obvious example of such a machine is the electric motor. Another example is the relay, an electrically-controlled switch. If a switch contact mechanism is built so that it can be actuated (opened and closed) by the application of a magnetic field, and an electromagnet coil is placed in the near vicinity to produce that requisite field, it will be possible to open and close the switch by the application of a current through the coil. In effect, this gives us a device that enables electricity to control electricity. See Fig. 2. 3.

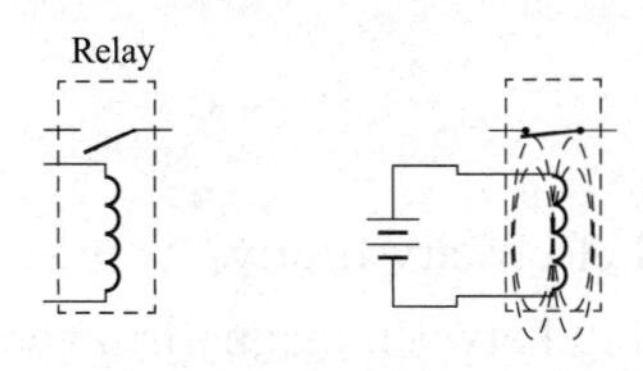

Fig. 2. 3 Applying current through the coil causes the switch to close

Words and Phrases

Magnetism *n.* 磁学
electricity *n.* 电学
stumble upon 偶然发现
Danish *adj.* 丹麦的
Oersted *n.* 奥斯特（人名）
one another 彼此，互相
compass needle 罗盘针
suspend *v.* 吊，悬挂，悬浮
compass *n.* 罗盘
electric current 电流
conjecture *n.* 推测，猜想
serendipitous *adj.* 偶然发现的
electromagnetics *n.* 电磁学
detailed experiment 详细的试验
magnetic field 磁场
perpendicular *adj.* 垂直的，正交的
right-hand rule 右手定则
encircle *v.* 环绕，围绕
magnetic flux line 磁力线
current-carrying wire 载流导线
magnetomotive force 磁动势
electromotive force 电动势
electric motor 电动机
electrically-controlled switch 电控开关

Notes

1. By passing an electric current through a metal wire suspended above a magnetic

compass, Oersted was able to produce a definite motion of the compass needle in response to the current.

奥斯特在悬放于磁罗盘上方的金属线中通入电流，发现罗盘针有明确的运动。

2. The right-hand rule says that the magnetic flux lines produced by a current-carrying wire will be oriented the same direction as the curled fingers of a person's right hand, with the thumb pointing in the direction of electron flow.

右手定则：当大拇指指向电流的方向时，载流导线所产生的磁力线的方向与人的右手四指的弯曲方向一致。

3. The amount of magnetic field force generated by a coiled wire is proportional to the current through the wire multiplied by the number of "turns" or "wraps" of wire in the coil.

导线线圈产生的磁动势的大小与导线中流过的电流和线圈匝数的乘积呈正比。

Exercises

1. Answer the following questions according to the text

(1) Who discovered the relationship between magnetism and electricity firstly?

(2) What is the right-hand rule used to explain the relationship between magnetism and electricity?

(3) The magnetic field force produced by a current-carrying wire can be greatly increased by shaping the wire into a coil instead of a straight line. Right or wrong?

(4) What is very much analogous to electromotive force (E) in an electric circuit?

(5) Where is Electromagnets widely used in?

2. Translate the following sentences into Chinese according to the text

(1) His serendipitous discovery paved the way for a whole new branch of science: electromagnetics.

(2) To create a stronger magnetic field force with the same amount of electric current, we can wrap the wire into a coil shape, where the circling magnetic fields around the wire will join to create a larger field with a definite magnetic polarity.

(3) An electromagnet is a piece of wire intended to generate a magnetic field with the passage of electric current through it.

(4) As an electrically-controllable magnet, electromagnets find application in a wide variety of "electromechanical" devices: machines that effect mechanical force or motion through electrical power.

3. Translate the following paragraph into Chinese

Electromechanical devices which employ magnetic fields often use ferromagnetic materials for guiding and concentrating these fields. Because the magnetic permeability of ferromagnetic materials can be large (up to tens of thousands times that of the surrounding space), most of the magnetic flux is confined to fairly well-defined paths determined by the geometry of the magnetic material.

Text B Permeability and Saturation

The nonlinearity of material permeability may be graphed for better understanding. We'll place the quantity of field intensity (H), equal to field force divided by the length of the material, on the horizontal axis of the graph. On the vertical axis, we'll place the quantity of flux density (B), equal to total flux divided by the cross-sectional area of the material. We will use the quantities of field intensity (H) and flux density (B) instead of field force and total flux (Φ) so that the shape of our graph remains independent of the physical dimensions of our test material. What we're trying to do here is show a mathematical relationship between field force and flux for any chunk of a particular substance.

Fig. 2. 4 shows the normal magnetization curve, or B-H curve. Notice how the flux density for any of the above materials (cast iron, cast steel, and sheet steel) levels off with increasing amounts of field intensity. This effect is known as saturation. When there is little applied magnetic force (low H), only a few atoms are in alignment, and the rest are easily aligned with additional force. However, as more flux gets crammed into the same cross-sectional area of a ferromagnetic material, fewer atoms are available within that material to align their electrons with additional force, and so it takes more and more force (H) to get less and less "help" from the material in creating more flux density (B). Saturation is a phenomenon limited to iron-core electromagnets. Air-core electromagnets don't saturate, but on the other hand they don't produce nearly as much magnetic flux as a ferromagnetic core for the same number of wire turns and current.

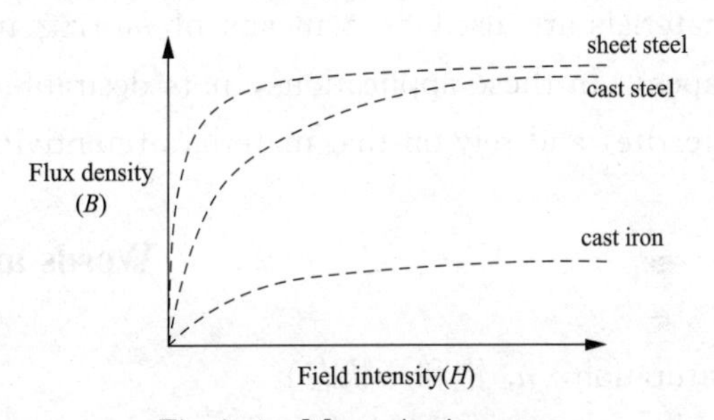

Fig. 2. 4 Magnetization curve

As a general term, hysteresis means a lag between input and output in a system upon a change in direction. In a magnetic system, hysteresis is seen in a ferromagnetic material that tends to stay magnetized after an applied field force has been removed.

Let's use the same graph again, only extending the axes to indicate both positive and negative quantities. First we'll apply an increasing field force (current through the coils of our electromagnet). We should see the flux density increase (go up and to the right) according to the normal magnetization curve, see Fig. 2. 5. Next, we'll stop the current going through the coil of the electromagnet and see what happens to the flux, leaving the first curve still on the graph. Due to the retentivity of the material, we still have a magnetic flux with no applied force (no current through the coil). Our electromagnet core is acting as a permanent magnet at this point. Now we will slowly apply the same amount of magnetic field force in the opposite direction to our sample. The flux density has now reached a point equivalent to what it was with a full positive value of field intensity (H), except in the negative, or opposite, direction. Let's stop the current going through

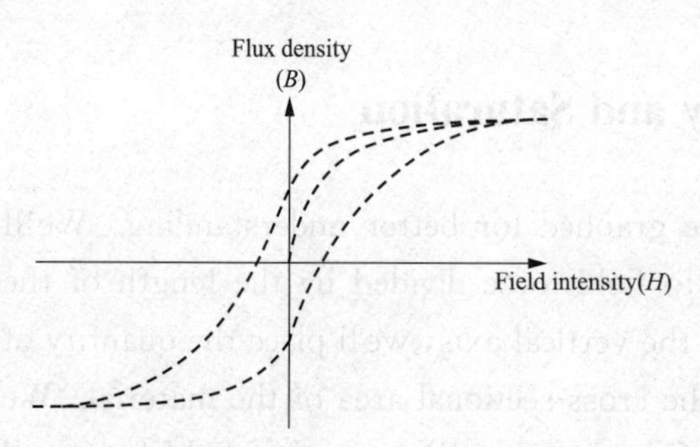

Fig. 2. 5 Hysteresis curve of a ferromagnetic material

the coil again and see how much flux remains. Once again, due to the natural retentivity of the material, it will hold a magnetic flux with no power applied to the coil, except this time it's in a direction opposite to that of the last time we stopped current through the coil. If we re-apply power in a positive direction again, we should see the flux density reach its prior peak in the upper-right corner of the graph again.

The "S"-shaped curve traced by these steps form what is called the hysteresis curve of a ferromagnetic material for a given set of field intensity extremes ($-H$ and $+H$).

Hysteresis can be a problem. If you're designing a system to produce precise amounts of magnetic field flux for given amounts of current, hysteresis may hinder this design goal (due to the fact that the amount of flux density would depend on the current and how strongly it was magnetized before). Having to overcome prior magnetization in an electromagnet can be a waste of energy if the current used to energize the coil is alternating. The area within the hysteresis curve gives a rough estimate of the amount of this wasted energy.

Other times, magnetic hysteresis is a desirable thing. Such is the case when magnetic materials are used as a means of storing information (computer disks, audio and video tapes). In these applications, it is desirable to be able to magnetize a speck of iron oxide (ferrite) and rely on that material's retentivity to "remember" its last magnetized state.

Words and Phrases

saturation *n.* 饱和（状态）
nonlinearity *n.* 非线性
material *n.* 材料
magnetic field intensity 磁场强度
magnetic field force 磁势
horizontal axis 横坐标
vertical axis 纵坐标
cross-sectional area 横截面积
ferromagnetic material 铁磁材料
normal magnetization curve 基本磁化曲线
level off 变平，稳定
cast *n.* 铸件；*v.* 浇铸
cast iron 铸铁
cast steel 铸钢
sheet steel 钢板
atom *n.* 原子
retentivity *n.* 矫顽力
lag *n.* 落后，迟延
negative *adj.* 负的
positive *adj.* 正的
opposite *adj.* 相反的，相对的
direction *n.* 方向
graph *n.* 曲线图
extreme *n.* 极值
rough *adj.* 粗略的，大致的
precise *adj.* 精确的
repeatable *adj.* 可重复的
upper-right corner 右上角
"S"-shaped curve S形曲线
hysteresis curve 滞回曲线，磁滞回线

Notes

1. We'll place the quantity of field intensity (H), equal to field force divided by the length of the material, on the horizontal axis of the graph.

我们用磁场强度即磁动势除以材料的长度作为图的横坐标。

2. Saturation is a phenomenon limited to iron-core electromagnets. Air-core electromagnets don't saturate, but on the other hand they don't produce nearly as much magnetic flux as a ferromagnetic core for the same number of wire turns and current.

铁心电磁铁才有饱和现象。空心电磁不会饱和，但另一方面，相同安匝数的磁动势在空心线圈中产生的磁通远小于在铁心线圈中产生的磁通。

Exercises

1. Answer the following questions according to the text

(1) What is the relation between field intensity (H) and field force?

(2) What is the relation between flux density (B) and total flux?

(3) What is saturation of ferromagnetic material?

(4) Saturation is a phenomenon limited to iron-core electromagnets. Air-core electromagnets don't saturate. Right or wrong?

(5) The area within the hysteresis curve gives a rough estimate of the amount of this wasted energy. Right or wrong?

2. Translate the following sentences into Chinese according to the text

(1) Notice how the flux density for any of the above materials (cast iron, cast steel, and sheet steel) levels off with increasing amounts of field intensity.

(2) Due to the retentivity of the material, we still have a magnetic flux with no applied force (no current through the coil).

(3) The flux density has now reached a point equivalent to what it was with a full positive value of field intensity (H), except in the negative, or opposite, direction.

(4) If we re-apply power in a positive direction again, we should see the flux density reach its prior peak in the upper-right corner of the graph again.

(5) Other times, magnetic hysteresis is a desirable thing. Such is the case when magnetic materials are used as a means of storing information.

3. Translate the following paragraph into Chinese

Due to the hysteresis effect, the relationship between B and H for a ferromagnetic material is both nonlinear and multivalued. In general, the characteristics of the material cannot be described analytically. They are commonly presented in graphical form as a set of empirically determined curves based on test samples of the material. The most common curve used to describe a magnetic material is the B-H curve or hysteresis loop.

Unit 2 Electric Network Analysis

Text A Branch Current Method

Generally speaking, network analysis is any structured technique used to mathematically analyze a circuit (a "network" of interconnected components). Quite often the technician or engineer will encounter circuits containing multiple sources of power or component configurations which defy simplification by series/parallel analysis techniques. In those cases, he or she will be forced to use other means. Branch Current Method and Mesh Current Method are useful techniques in analyzing such complex circuits.

The first and most straightforward network analysis technique is called the Branch Current Method. In this method, we assume directions of currents in a network, then write equations describing their relationships to each other through Kirchhoff's and Ohm's Laws. Once we have one equation for every unknown current, we can solve the simultaneous equations and determine all currents, and therefore all voltage drops in the network.

Let's use this circuit, shown in Fig. 2.6, to illustrate this method.

The first step is to choose a node (junction of wires) in the circuit to use as a point of reference for our unknown currents. I'll choose the node joining the right of R_1, the top of R_2, and the left of R_3. See Fig. 2.7.

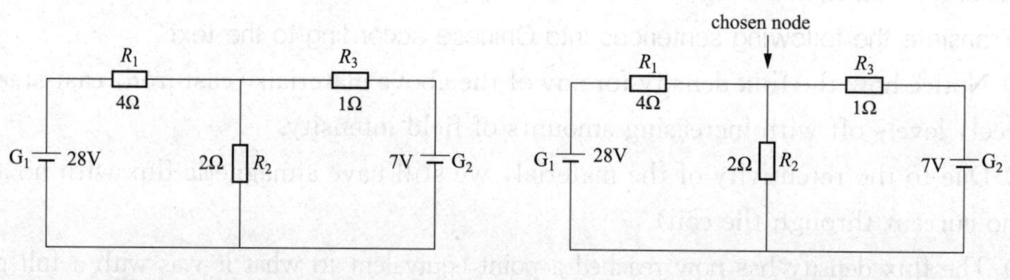

Fig. 2.6 Circuit example Fig. 2.7 Chosen node

At this node, guess which directions the three wires' currents take, labeling the three currents as I_1, I_2, and I_3, respectively. See Fig. 2.8.

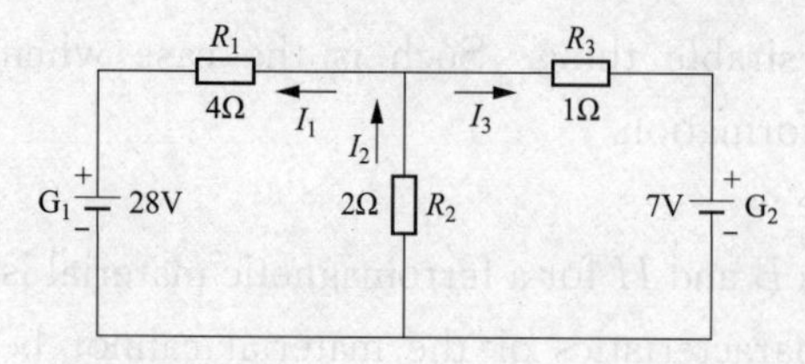

Fig. 2.8 Labeling the currents

Bear in mind that these directions of current are speculative at this point. Fortunately, if it turns out that any of our guesses were wrong, we will know when we mathematically solve for the currents (any "wrong" current directions will show up as negative numbers in our solution).

Kirchhoff's Current Law (KCL) tells us that the algebraic sum of currents entering and exiting a node must equal zero, so we can relate these three currents (I_1, I_2, and I_3) to each other in a single equation. For the sake of convention, I'll denote any current entering the node as positive in sign, and any current exiting the node as negative in sign: $-I_1+I_2-I_3=0$

The next step is to label all voltage drop polarities across resistors according to the assumed directions of the currents. Remember that the "upstream" end of a resistor will always be negative, and the "downstream" end of a resistor positive with respect to each other, since electrons are negatively charged. See Fig. 2. 9.

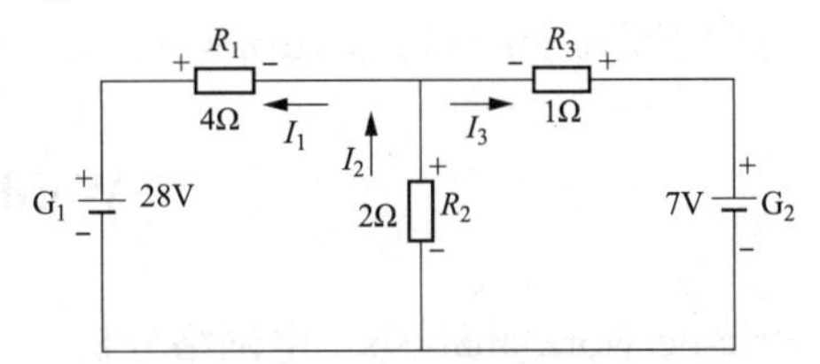

Fig. 2. 9 Labeling the voltage drop polarities

The battery polarities, of course, remain as they were according to their symbology (short end negative, long end positive). It is okay if the polarity of a resistor's voltage drop doesn't match with the polarity of the nearest battery, so long as the resistor voltage polarity is correctly based on the assumed direction of current through it. In some cases we may discover that current will be forced backwards through a battery, causing this very effect. The important thing to remember here is to base all your resistor polarities and subsequent calculations on the directions of current (s) initially assumed. As stated earlier, if your assumption happens to be incorrect, it will be apparent once the equations have been solved (by means of a negative solution). The magnitude of the solution, however, will still be correct.

Kirchhoff's Voltage Law (KVL) tells us that the algebraic sum of all voltages in a loop must equal zero, so we can create more equations with current terms (I_1, I_2, and I_3) for our simultaneous equations. To obtain a KVL equation, we must tally voltage drops in a loop of the circuit, as though we were measuring with a real voltmeter. I'll choose to trace the left loop of this circuit first, starting from the upper-left corner and moving counter-clockwise (the choice of starting points and directions is arbitrary).

Notice how current is being pushed backwards through battery G_2 (electrons flowing "up") due to the higher voltage of battery G_1 (whose current is pointed "down" as it normally would)! Despite the fact that battery G_2's polarity is trying to push electrons down in that branch of the circuit, electrons are being forced backwards through it due to the superior voltage of battery G_1. Does this mean that the stronger battery will always "win" and the weaker battery always get current forced through it backwards? No! It actually depends on both the batteries' relative voltages and the resistor values in the circuit. The only sure way to determine what's going on is to take the time to mathematically analyze the network.

Now that we know the magnitude of all currents in this circuit, we can calculate voltage drops across all resistors with Ohm's Law ($E=IR$).

Steps to follow for the "Branch Current" method of analysis:

(1) Choose a node and assume directions of currents.

(2) Write a KCL equation relating currents at the node.

(3) Label resistor voltage drop polarities based on assumed currents.

(4) Write KVL equations for each loop of the circuit, substituting the product IR for E in each resistor term of the equations.

(5) Solve for unknown branch currents (simultaneous equations).

(6) If any solution is negative, then the assumed direction of current for that solution is wrong!

(7) Solve for voltage drops across all resistors ($E=IR$).

Words and Phrases

electric network analysis 电网络分析
structured technique 构造技术（方法）
interconnected *adj*. 互相联系的
component *n*. 组成部分
technician *n*. 技术员，技师
Branch Current Method 支路电流法
Mesh Current Method 网孔电流法
equation *n*. 等式，方程式
Kirchhoff's Law 基尔霍夫定律
Ohm's Law 欧姆定律
simultaneous equations 联立方程式
node *n*. 节点
label *v*. 标注；*n*. 标签
bear in mind 记住
Kirchhoff's Current Law (KCL) 基尔霍夫电流定律
algebraic sum 代数和
polarity *n*. 极性
symbology *n*. 象征学，符号
arbitrary *adj*. 任意的
battery *n*. 电池
electron *n*. 电子
magnitude *n*. 数量，大小
Kirchhoff's Voltage Law (KVL) 基尔霍夫电压定律
tally *v*. 计算，记录，加标签于
loop *n*. 回路
voltmeter *n*. 电压表，伏特计
clockwise *adj*. 顺时针方向的
resistor *n*. 电阻器

Notes

1. Quite often the technician or engineer will encounter circuits containing multiple sources of power or component configurations which defy simplification by series/parallel analysis techniques.

技术员或工程师常常会遇到一些包含若干电源或其他组成部分，并且难于用串并联分析方法将其简化的电路。

2. For the sake of convention, I'll denote any current entering the node as positive in sign, and any current exiting the node as negative in sign.

按惯例，将流入节点的电流记为正，流出节点的电流记为负。

Exercises

1. Answer the following questions according to the text

(1) What is electric network analysis?

(2) Which Methods are used in analyzing complex circuits?

(3) What is the first step when using Branch Current Method to analyze a complex circuits?

(4) In one circuit, the stronger battery will always "win" and the weaker battery always get current forced through it backwards. Right or wrong?

2. Translate the following sentences into Chinese according to the text

(1) In Branch Current Method, we assume directions of currents in a network, then write equations describing their relationships to each other through Kirchhoff's and Ohm's Laws.

(2) Kirchhoff's Current Law (KCL) tells us that the algebraic sum of currents entering and exiting a node must equal zero, so we can relate these three currents (I_1, I_2, and I_3) to each other in a single equation.

(3) The next step is to label all voltage drop polarities across resistors according to the assumed directions of the currents.

(4) It is okay if the polarity of a resistor's voltage drop doesn't match with the polarity of the nearest battery, so long as the resistor voltage polarity is correctly based on the assumed direction of current through it.

(5) Kirchhoff's Voltage Law (KVL) tells us that the algebraic sum of all voltages in a loop must equal zero.

3. Translate the following paragraph into Chinese

The Mesh Current Method is quite similar to the Branch Current method in that it uses simultaneous equations, Kirchhoff's Voltage Law, and Ohm's Law to determine the unknown currents in a network. It differs from the Branch Current method in that it does not use Kirchhoff's Current Law, and it is usually able to solve a circuit with less unknown variables and less simultaneous equations, which is especially nice if you're forced to solve without a calculator.

Text B Network Theorems

In electric network analysis, the fundamental rules are Ohm's Law and Kirchhoff's Laws. While these humble laws may be applied to analyze just about any circuit configuration (even if we have to resort to complex algebra to handle multiple unknowns), there are some "shortcut" methods of analysis to make the math easier for the technician or engineer.

As with any theorem of geometry or algebra, these network theorems, such as Norton's Theorem, Superposition Theorem, Thevenin's Theorem, etc., are derived from fundamental rules.

In Millman's Theorem, the circuit is re-drawn as a parallel network of branches, each branch containing a resistor or series battery/resistor combination. Millman's Theorem is applicable only to those circuits which can be re-drawn accordingly.

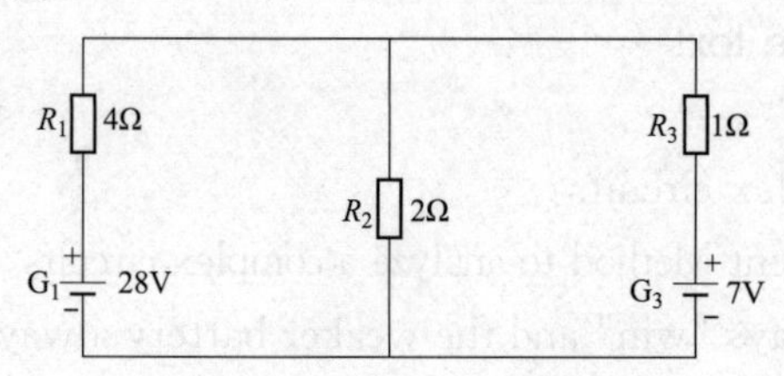

Fig. 2. 10 The re-drawn of Fig. 2. 6

The circuit of Fig. 2. 6 is re-drawn as Fig. 2. 10 for the sake of applying Millman's Theorem.

By considering the supply voltage within each branch and the resistance within each branch, Millman's Theorem will tell us the voltage across all branches. Please note that I've labeled the battery in the rightmost branch as "G_3" to clearly denote it as being in the third branch.

Millman's Theorem is nothing more than a long equation, applied to any circuit drawn as a set of parallel-connected branches, each branch with its own voltage source and series resistance:

$$\frac{\frac{E_{B1}}{R_1}+\frac{E_{B2}}{R_2}+\frac{E_{B3}}{R_3}}{\frac{1}{R_1}+\frac{1}{R_2}+\frac{1}{R_3}}=\text{Voltage across all branches}$$

Substituting actual voltage and resistance figures from our example circuit for the variable terms of this equation, we get the following expression:

$$\frac{\frac{28}{4}+\frac{0}{2}+\frac{7}{1}}{\frac{1}{4}+\frac{1}{2}+\frac{1}{1}}=8(\text{V})$$

The final answer of 8 V is the voltage seen across all parallel branches, like Fig. 2. 11.

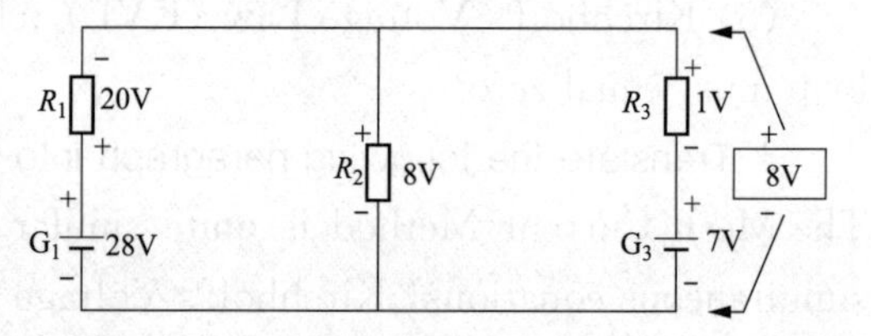

Fig. 2. 11 Voltage across all parallel branches

The polarity of all voltages in Millman's Theorem are referenced to the same point. In the example circuit above, we used the bottom wire of the parallel circuit as my reference point, and so the voltages within each branch (28V for the R_1 branch, 0V for the R_2 branch, and 7V for the R_3 branch) were inserted into the equation as positive numbers. Likewise, when the answer came out to 8 V (positive), this meant that the top wire of the circuit was positive with respect to the bottom wire (the original point of reference). If both batteries had been connected backwards (negative ends up and positive ends down), the voltage for branch 1 would have been entered into the equation as a −28 V, the voltage for branch 3 as −7 V, and the resulting answer of −8 V would have told us that the top wire was negative with respect to the bottom wire (our initial point of reference).

To solve for resistor voltage drops, the Millman voltage (across the parallel network) must be compared against the voltage source within each branch, using the principle of voltages adding in series to determine the magnitude and polarity of voltage across each resistor:

$$E_{R1} = 8 - 28 = -20(\text{V})(\text{negative on top})$$
$$E_{R2} = 8 - 0 = 8(\text{V})(\text{positive on top})$$
$$E_{R3} = 8 - 7 = 1(\text{V})(\text{positive on top})$$

To solve for branch currents, each resistor voltage drop can be divided by its respective resistance ($I=E/R$):

$$I_{R1} = \frac{20}{4} = 5(\text{A})$$
$$I_{R2} = \frac{8}{2} = 4(\text{A})$$
$$I_{R3} = \frac{1}{1} = 1(\text{A})$$

The direction of current through each resistor is determined by the polarity across each resistor, not by the polarity across each battery, as current can be forced backwards through a battery, as is the case with G_3 in the example circuit. See Fig. 2. 12. This is important to keep in mind, since Millman's Theorem doesn't provide as direct an indication of "wrong" current direction as does the Branch Current or Mesh Current methods. You must pay close attention to the polarities of resistor voltage drops as given by Kirchhoff's Voltage Law, determining direction of currents from that.

Millman's Theorem is very convenient for determining the voltage across a set of parallel branches, where there are enough voltage sources present to preclude solution via regular series-parallel reduction method. It also is easy in the sense that it doesn't require the use of simultaneous equations. However, it is limited in that it only applied to circuits which can be re-drawn to fit this form. It cannot be used, for example, to solve an unbalanced bridge circuit. And, even in cases where Millman's Theorem can be applied, the solution of individual resistor voltage drops can be a bit daunting to some, the Millman's Theorem equation only providing a single figure for branch voltage.

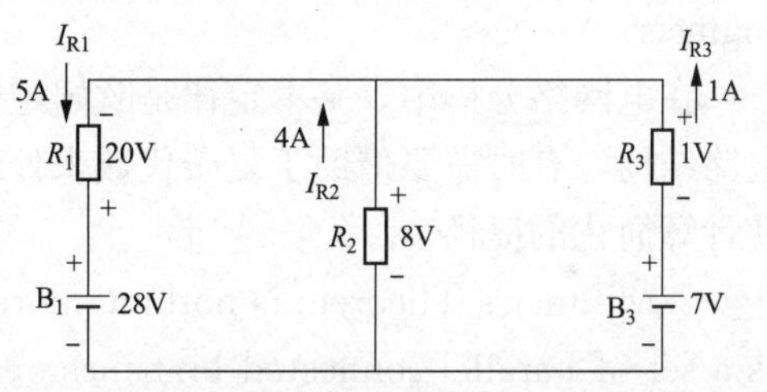

Fig. 2. 12 The direction of current through each resistor

Each network analysis method has its own advantages and disadvantages. Each method is a tool, and there is no tool that is perfect for all jobs. The skilled technician, however, carries these methods in his or her mind like a mechanic carries a set of tools in his or her tool box. The more tools you have equipped yourself with, the better prepared you will be for any eventuality.

Words and Phrases

humble *adj.* 粗陋的，谦逊的

handle *v.* 处理，操作

shortcut *n.* 捷径
theorem *n.* 定理，法则
geometry *n.* 几何学
algebra *n.* 代数学
branch *n.* 支路
rightmost *adj.* 最右边的
parallel-connected branch 并联支路
series resistance 串联电阻
negative ends up and positive ends down 负端在上，正端在下
voltage source 电压源
likewise *adv.* 同样地
resistor voltage drop 电阻压降
branch current 支路电流
convenient *adj.* 方便的
determine *v.* 决定，确定
preclude *v.* 排除
eventuality *n.* 可能性
equip *v.* 装备，准备
advantage and disadvantage 优点和缺点
skilled technician 熟练的技术员

Notes

1. In electric network analysis, the fundamental rules are Ohm's Law and Kirchhoff's Laws. While these humble laws may be applied to analyze just about any circuit configuration (even if we have to resort to complex algebra to handle multiple unknowns), there are some "shortcut" methods of analysis to make the math easier for the technician or engineer.

在电网络分析中，基本定律是欧姆定律和基尔霍夫定律，这些简单定律适用于分析任何电路结构，有时需要借助于复杂代数来处理多个未知量。这里向工程技术人员介绍一些可以使计算简化的捷径。

2. Millman's Theorem is nothing more than a long equation, applied to any circuit drawn as a set of parallel-connected branches, each branch with its own voltage source and series resistance.

米尔曼（Millman）定律就是由各并联支路组成的一个长等式，其中每条支路各有其电压源和串联电阻。

3. Millman's Theorem is very convenient for determining the voltage across a set of parallel branches, where there are enough voltage sources present to preclude solution via regular series-parallel reduction method.

米尔曼定律可以很方便地确定各并联支路的电压，通过串并联化简方法各支路都可以有电压源。

Exercises

1. Answer the following questions according to the text

(1) How many electric network theorems do you know? What are they?

(2) Millman's Theorem is applicable to any circuits. Right or wrong?

(3) The polarity of all voltages in Millman's Theorem is referenced to the same point.

Right or wrong?

(4) The direction of current through each resistor is determined by the polarity across each resistor, not by the polarity across each battery. Right or wrong?

(5) Each network analysis method has its own advantages and disadvantages. Each method is a tool, and there is no tool that is perfect for all jobs. Right or wrong?

2. Translate the following sentences into Chinese according to the text

(1) In Millman's Theorem, the circuit is re-drawn as a parallel network of branches, each branch containing a resistor or series battery/resistor combination.

(2) By considering the supply voltage within each branch and the resistance within each branch, Millman's Theorem will tell us the voltage across all branches.

(3) To solve for resistor voltage drops, the Millman voltage (across the parallel network) must be compared against the voltage source within each branch.

(4) To solve for branch currents, each resistor voltage drop can be divided by its respective resistance.

(5) This is important to keep in mind, since Millman's Theorem doesn't provide as direct an indication of "wrong" current direction as does the Branch Current or Mesh Current methods.

3. Translate the following paragraph into Chinese

The strategy used in the Superposition Theorem is to eliminate all but one source of power within a network at a time, using series/parallel analysis to determine voltage drops (and/or currents) within the modified network for each power source separately. Then, once voltage drops and/or currents have been determined for each power source working separately, the values are all "superimposed" on top of each other to find the actual voltage drops/currents with all sources active.

Unit 3 Computation of Electromagnetic Fields

Text A Finite Element Methods for Electromagnetic Field Simulation

Here are two attributes of the method of finite elements that have prompted the rapid growth of its application to the modeling of electromagnetic interactions. One of them is its superior modeling versatility where structures of arbitrary shape and composition can be modelled as precisely as the desirable model complexity and available computer resources dictate. The second, is common to all differential equation-based numerical methods, and has to do with the fact that the matrix resulting from the discretization of the governing equations is very sparse, which implies savings in computer memory for its storage as well as in CPU time for its inversion. Clearly, these two attributes come at the expense of an increase in the degrees of freedom used in the numerical approximation of the problem since now, contrary to integral equation methods, the entire space surrounding all sources of electromagnetic fields needs be incorporated in the numerical model. Nevertheless, because of the sparsity of the resulting matrix and the simplicity with which complex geometries can be modeled, this increase in the degrees of freedom of the approximation is an acceptable penalty.

1. Mathematical Framework for Finite Element Analysis

In order to review the basic steps involved in the finite element approximation of electromagnetic boundary value problems, let us consider the double-curl equation for the electric field, $\boldsymbol{E}$, which, in a source-free, isotropic and linear medium with position-dependent magnetic and electric properties has the form

$$\nabla\times\left(\frac{1}{\mathrm{j}\omega\mu}\nabla\times\boldsymbol{E}\right)+\mathrm{j}\omega\hat{\varepsilon}\boldsymbol{E}=0 \tag{2-1}$$

The time dependence $e^{j\omega t}$ is assumed ($j=\sqrt{-1}$), and the complex permittivity, $\hat{\varepsilon}=\varepsilon-j\sigma/\omega$, is used to account for any conduction and/or dielectric losses in the medium. For the purposes of finite element solutions, a weak form of Equation (2-1) is required. For node based finite element expansions the unknown vector field is approximated in terms of scalar basis functions, ϕ_i

$$E=\sum_i \boldsymbol{E}_i\phi_i \tag{2-2}$$

where E_i denotes the unknown vector field value at node i. The relevant weak form, in the spirit of Galerkins approximation, is

$$\left\langle\left(\frac{1}{\mathrm{j}\omega\mu}\nabla\times\boldsymbol{E}\right)\times\nabla\phi_i\right\rangle+\left\langle\mathrm{j}\omega\hat{\varepsilon}\boldsymbol{E}\phi_i\right\rangle=-\oint\frac{1}{\mathrm{j}\omega\mu}\hat{n}\times(\nabla\times\boldsymbol{E})\phi_i\,\mathrm{d}s \tag{2-3}$$

where $\langle\ \rangle$ and $\oint$ indicate integration over the domain of interest and its boundary, respectively, while $\hat{n}$ is the outward unit normal on the boundary.

The most well-known attribute of Galerkins method, where the solution is sought in a finite-dimensional subspace of the class of admissible functions for the problem of interest using the same set of functions as trial and test functions, is the symmetry of the resulting stiffness matrix given a symmetric weak formulation. However, another important merit of Galerkins method is that, if a symmetric weak formulation is used, Galerkins approximate solution exactly conserves energy in the electromagnetic field despite the fact that it satisfies the vector Helmholtz equation only approximately over the domain of interest.

2. Grid Generation

Numerical grid generation is probably the most critical step in a finite element analysis of electromagnetic wave interactions. During the early stages of the application of the finite element method to modeling of electromagnetic interactions, the emphasis was on mathematical model and weak statement formulations and their subsequent use in the analysis of propagation, radiation and scattering problems in conjunction with rather simple geometries.

Consequently, the important issue of automatic generation of finite element grids appropriate for electromagnetic propagation and scattering problems received rather limited attention. Apparently, the assumption was that grid generation practices used in low-frequency electromagnetic field modeling and/or other areas of engineering in which the method of finite elements was already being used, could be adopted without significant alterations. This turned out to be a rather false assumption, simply because the specific physics of the phenomenon that is being analyzed needs be taken into account before a discrete model is built for its quantification. The famous sampling theorem of modern communication theory for the sampling of a band-limited signal serves as a simple, yet powerful example.

While it is often tempting to adopt the philosophy that the finer the grid the better the quality of the solution, one needs to remember that an excessively fine mesh (unless needed) wastes computational resources and thus it should be avoided. Consequently, it is important that the grid generation process is such that adaptive mesh refinement is possible.

3. The Choice of Elements

Most of the original applications of the finite element method to vectorial electromagnetic field modeling in three dimensions were based on the so-called nodal elements, using field representations of the form shown in Equation (2-2) and weak statements such as the one in Equation (2-3). In other words, the degrees of freedom were defined to be the three components of the unknown field quantity at the element nodes. It was soon found that such approximations were plagued by the occurrence of non-physical, spurious modes. These spurious solutions manifested themselves as modes with nonzero divergence, and were caused by the inability of the aforementioned choice of interpolation and weak statement to enforce Gauss's law for divergence-free solutions. Consequently, a variety of approaches were proposed for eliminating these spurious modes within the context of nodal elements.

The most popular version of these edge elements is the so-called Whitney I-form. It was long before the method of finite elements was becoming a popular tool in boundary value problem solving that Whitney described a family of polynomial forms on a simplicial mesh with special properties that made them attractive for electromagnetic field representations.

The above list is expected to grow substantially as computing technology continues to advance rapidly, providing us with higher computation speeds, larger memory resources, parallelism and distributed computing. Continuing research in advancing the state-of-the-art in automatic grid generation and refinement, improving the performance and robustness of reflectionless grid truncation, and automating the application of domain decomposition approaches, will help enhance the power of the finite element method to solve realistic engineering problems.

Words and Phrase

discretization *n.* 离散化

sparsity *n.* 稀疏

integral equation 积分方程

penalty *n.* 惩罚，处罚，害处，不利

curl *n.* 旋度

isotropic *adj.* 各向同性的

stiffness matrix 刚度矩阵

trial function 试探函数

propagation *n.* 传播

radiation *n.* 辐射

scattering *n.* 散射

plague *v.* 困扰，折磨，使煎熬

spurious *adj.* 虚假的，伪造的

distributed computing 分布式计算

truncation *n.* 截断

Notes

1. Nevertheless, because of the sparsity of the resulting matrix and the simplicity with which complex geometries can be modeled, this increase in the degrees of freedom of the approximation is an acceptable penalty.

然而，由于所得到的矩阵的稀疏性和复杂几何形状的简单性均可模拟，因此近似的自由度的增加是可接受的补偿。

2. The most well-known attribute of Galerkins method, where the solution is sought in a finite-dimensional subspace of the class of admissible functions for the problem of interest using the same set of functions as trial and test functions, is the symmetry of the resulting stiffness matrix given a symmetric weak formulation.

伽辽金方法中最著名的属性是：给定一个对称的弱公式，由此得到的刚度矩阵的对称性，该方法使用与试探函数和检验函数相同的一组函数，在一类容许函数的有限维子空间中寻求感兴趣问题的解。

3. In other words, the degrees of freedom were defined to be the three components of the unknown field quantity at the element nodes.

换言之，自由度被定义为元素节点处未知场量的三个分量。

Exercises

1. Answer the following questions according to the text

(1) How many attributes of the finite element method?

(2) What are the merits of the Galerkins method?

(3) It is important that the grid generation process is such that adaptive mesh refinement is possible. Right or wrong?

(4) How to define the so-called nodal elements.

2. Translate the following sentences into Chinese according to the text

(1) For node based finite element expansions the unknown vector field is approximated in terms of scalar basis functions.

(2) Numerical grid generation is probably the most critical step in a finite element analysis of electromagnetic wave interactions.

(3) The famous sampling theorem of modern communication theory for the sampling of a band-limited signal serves as a simple, yet powerful example.

(4) Consequently, it is important that the grid generation process is such that adaptive mesh refinement is possible.

3. Translate the following paragraph into Chinese

The finite element procedure consists of partitioning or discretizing interior domain Ω into a number of subdomains or finite elements. The field is approximated over each element by an interpolating or shape function depending on values at discrete nodes on or in the element. To provide some degree of field continuity across element boundaries, most of the discrete nodes are defined on the element surfaces and shared by adjacent elements. Provided that they cover the domain, the elements and shape functions may be completely arbitrary.

Text B Time Domain versus Frequency Domain

Why does EMflex solve steady-state problems with a time-domain algorithm? This is an obvious, fair, and often asked question. The short answer is that time-domain methods currently provide the quickest, least computer memory intensive, most robust path to a solution. Advances in numerical methods may someday change this answer. There do exist reasons for using frequency-domain simulations if they become competitive in CPU and memory cost. In this section, we outline the issues governing the choice between time-domoin and frequency-domain solvers for finite element models, list the pros and cons of each approach, present some comparisons between EMflex and an equivalent frequency domain formulation for the 2D scalar case (using state-of-the-art iterative solvers), and conclude with some observations on the future outlook for frequency-domain computations.

The primary advantage of the time-domain solver used in EMflex is that it embodies an efficient explicit algorithm that requires minimal memory, thus permitting solutions of the largest finite element model possible on any given machine. It is also robust and deterministic in the sense that if run long enough, steady-state will be achieved and a solution will be found. Additional advantages of the time-domain approach are that it is directly extensible to nonlinear problems where material properties change with time.

Disadvantages of time-domain solvers are that user intervention is required to assess when steady-state has been reached. One can envision problems where a long simulation may be necessary to achieve steady-state. Also, the steady-state quantities (amplitude and phase) are not primary in the time domain, and must be obtained by a secondary calculation after steady-state has been achieved. Note that this process has been automated inmerle.

Given that we are looking for a frequency-domain solution, the advantages of a straightforward frequency-domain formulation appear obvious. The difficulty is that for realistically sized models this formulation requires the solution of an enormous, sparse, linear system of equations. In addition, the linear system has undesirable numerical properties, namely, it is complex, non-hermitian (due to absorbing boundary conditions and/or conductivity), non-symmetric (due to absorbing boundary conditions), and is typically indefinite.

There are two basic methods of solving linear systems: direct (some variation of Gaussian elimination) and iterative. Direct methods are typically divided into two phases: factorization of the coefficient matrix, followed by back-substitution to obtain the solution. The drawback of direct methods is that they require large amounts of memory and CPU time. For example, consider a $100 \times 100 \times 100$ element 3D model with 3 electric field components unknown at each node, in complex arithmetic. This is a modest model size. For this problem, a standard band solver requires 5.73×10^{11} words of storage, which far exceeds the capacity of any presented computer. The number of floating point operations to solve such a system is also prohibitive. It is worth noting here that this linear system is very sparse, i. e. , only 2.5×10^{8} or 0.04% of the entries are nonzero. Observe that even the nonzero entries exceed the largest available machines, but only by a small margin. Some work has been done on sparse system direct solvers which attempt to economize on storage relative to the basic band solvers, but our experience with one such routine was disappointing. Thus we concur with the conventional wisdom which holds that 3D problems cannot be solved directly.

The alternative to direct solvers is iterative methods in which an initial guess for the solution (zero if nothing better is available) is successively refined until the error becomes "small". Conventional wisdom says that the Krylov subspace-type methods are best, which includes the conjugate gradient approach. Frequently, some type of preconditioner is used in conjunction with these methods to accelerate convergence. For some classes of matrices, e. g. ,positive definite symmetric, these methods work extremely well. The major portion of the CPU effort is in computing a product of the coefficient matrix with a vector and in computing inner products of two vectors. Thus, the insurmountable memory requirements of the direct methods can be avoided. Additional vectors of length N (the total number of unknowns), which form the basis of the Krylov Subspace, are usually required. For some iterative methods such as GMRES, these add up to many times more memory than for the time-domain algorithm.

Things to consider in choosing an iterative method are: ① the total amount of work required by the method and the preconditioner to achieve an acceptable solution; ② the total amount of memory required by the method and the preconditioner; ③ the possibility of breakdown—the algorithm fails, or hangs—the convergence rate becomes so low that a solution is not attainable.

For frequency-domain finite element solvers, we conclude that solution techniques for the resulting class of linear systems are still in the research phase. Significant advances have been achieved in recent years, and more advances are expected. We believe that trying new solution algorithms is a worthwhile exercise which provides feedback to the mathematicians and computer scientists, and is to everyone's advantage. On the other hand, it is premature to promote such solvers as production level tools for engineers. We conclude that if large-scale calculations need to be done today, time-domain techniques provide the most practical means of doing them.

Words and Phrases

steady-state *n.* 稳态
time-domain *n.* 时域
frequency-domain *n.* 频域
algorithm *n.* 算法
simulation *n.* 仿真
pro *n.* 正面，赞成
con *n.* 反面，反对
equivalent *adj.* 等价的，相当的
state-of-the-art *adj.* 艺术级的
iterative *adj.* 迭代的
nonlinear problem 非线性问题
material property 材料特性
intervention *n.* 干预，干涉
amplitude and phase 幅值和相位
sparse *adj.* 稀疏的
hermitian *adj.*（矩阵的）厄密共轭
indefinite *adj.* 模糊的，不确定的
phase *n.* 相角
factorization *n.* 因数分解
Gaussian elimination 高斯消去法
coefficient matrix 系数矩阵
substitution *n.* 代入法，取代
subspace *n.* 子空间
alternative *n.* 替代物
conjugate gradient approach 共轭梯度法
positive definite symmetric 正定对称的
preconditioner *v.* 预处理，前处理
accelerate convergence 加速收敛
matrix *n.* 矩阵
insurmountable *adj.* 不能克服的，不能超越的
advance *n.* 前进，提升
research phase 研究阶段

Notes

1. In addition, the linear system has numerical properties, namely, it is complex, non-hermitian (due to absorbing boundary conditions and/or conductivity), non-symmetric (due to absorbing boundary conditions), and is typically indefinite.

另外，线性系统有一些不良的数值特性，如：复杂、非共轭（由于吸收边界条件和/或传导性）、非对称（由于吸收边界条件）和典型的不确定性。

2. For example, consider a 100 × 100 × 100 element 3D model with 3 electric field components unknown at each node, in complex arithmetic. This is a modest model size. For this problem, a standard band solver requires 5.73×10^{11} words of storage, which far exceeds the capacity of any presented computer.

例如，用综合算法求解有100×100×100个单元、三维、每个节点3个电场未知量一个模型，这是一个大小适中的模型，此问题的求解需要5.73×10^{11}个字的存储量，这超出了任何一台现有计算机的计算能力。

Exercises

1. Answer the following questions according to the text

(1) Why does EMflex solve steady-state problems with a time-domain algorithm?

(2) What is the primary advantage of the time-domain solver used in EMflex?

(3) What are the disadvantages of the time-domain solver?

(4) What are the two basic methods of solving linear systems?

(5) If large-scale calculations need to be done today, time-domain techniques provide the most practical means of doing them. Right or wrong?

2. Translate the following sentences into Chinese according to the text

(1) The short answer is that time-domain methods currently provide the quickest, least computer memory intensive, most robust path to a solution. Advances in numerical methods may someday change this answer.

(2) In this section, we outline the issues governing the choice between time-domain and frequency-domain solvers for finite element model, list the pros and cons of each approach.

(3) The primary advantage of the time-domain solver used in EMflex is that it embodies an efficient explicit algorithm that requires minimal memory, thus permitting solutions of the largest finite element model possible on any given machine.

(4) Disadvantages of time-domain solvers are that user intervention is required to assess when steady-state has been reached.

(5) Things to consider in choosing an iterative method are: ① the total amount of work required by the method and the preconditioner to achieve an acceptable solution; ② the total amount of memory required by the method and the preconditioner; ③ the possibility of

breakdown—the algorithm fails, or hangs—the convergence rate becomes so low that a solution is not attainable.

3. Translate the following paragraph into Chinese

Discrete numerical methods like finite elements or finite differences are necessarily formulated on finite spatial grids—whether the actual domain being modeled is finite or infinite. This domain truncation introduces artificial boundaries that must be treated with special care in order to minimize nonphysical wave reflections. These trap energy that would otherwise be radiated and establish undesirable resonances within the grid. Note that this is true regardless of the solution scheme applied, in either the time-domain or frequency-domain.

Unit 4 Electromagnetic Interference and Electromagnetic Compatibility

Text A Electromagnetic Interference and Electromagnetic Compatibility

Electromagnetic interference (EMI) is a major problem in modern electronic circuits. To overcome the interference, the designer has to either remove the source of the interference, or protect the circuit being affected. The ultimate goal is to have the circuit operating as intended to achieve electromagnetic compatibility (EMC).

EMI: Electromagnetic emissions from a device or system that interfere with the normal operation of another device or system.

EMC: The ability of a device or system to function without error in its intended electromagnetic environment.

Examples of EMC problems:

- The use of mobile phones on the plane will interfere with the unlimited electrical contact between the plane and the ground control tower.
- Your car radio buzzes when you drive under a power line.
- A helicopter goes out of control when it flies too close to a radio tower.
- Your telephone is damaged by lightning-induced surges on the phone line.
- Your new memory board is destroyed by an unseen discharge as you install it.
- Noise of switching power supply.
- Your laptop computer interferes with your aircraft's rudder control.
- The airport radar interferes with your laptop computer display.

There are three essential elements to any EMC problem, which are shown in Fig. 2. 13. There must be a source of an electromagnetic phenomenon, a receptor (or victim) that cannot function properly due to the electromagnetic phenomenon, and a path between them that allows the source to interfere with the receptor. Each of these three elements must be present although they may not be readily identified in every situation. Electromagnetic compatibility problems are generally solved by identifying at least two of these elements and eliminating (or attenuating) one of them.

For example, in the case of the nuclear power plant, the receptor was readily identified. The turbine control valves were malfunctioning. The source and the coupling path were originally unknown, however an investigation revealed that the walkie talkies used by the plant employees were the source. Although at this point the coupling path was not known, the problem could be solved by eliminating the source (e. g. restricting the use of low power radio transmitters in certain areas). A more thorough and perhaps more secure approach

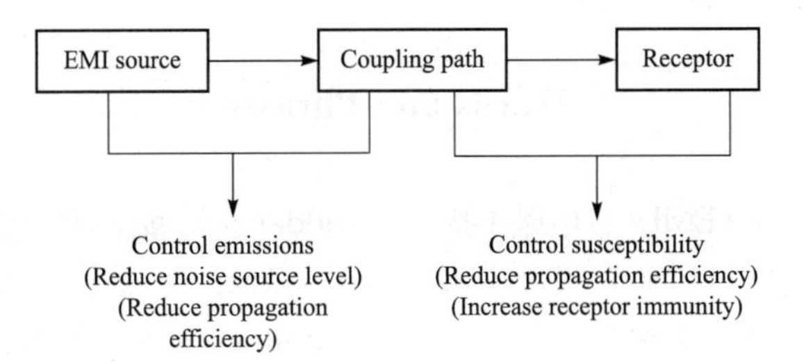

Fig. 2. 13 EMC elements

would be to identify the coupling path and take steps to eliminate it. For example, suppose it was determined that radiated emissions from a walkie talkie were inducing currents on a cable that was connected to a printed circuit card that contained a circuit that controlled the turbine valves. If the operation of the circuit was found to be adversely affected by these induced currents, a possible coupling path would be identified. Shielding, filtering, or rerouting the cable, and filtering or redesigning the circuit would then be possible methods of attenuating the coupling path to the point where the problem is non-existent.

The source of the tramway problem was thought to be transients on the tramway's power. The coupling path was presumably through the power supply to the speed control circuit, although investigators were unable to reproduce the failure so the source and coupling path were never identified conclusively. The receptor, on the other hand, was clearly shown to be the speed control circuit and this circuit was modified to keep it from becoming confused by unintentional random inputs. In other words, the solution was to eliminate the receptor by making the speed control circuit immune to the electromagnetic phenomenon produced by the source.

Potential sources of electromagnetic compatibility problems include radio transmitters, power lines, electronic circuits, lightning, lamp dimmers, electric motors, arc welders and just about anything that utilizes or creates electromagnetic energy. Potential receptors include radio receivers, electronic circuits, appliances, people, and just about anything that utilizes or can detect electromagnetic energy.

Methods of coupling electromagnetic energy from a source to a receptor fall into one of four categories.

➢ Conducted (electric current)

➢ Inductively coupled (magnetic field)

➢ Capacitively coupled (electric field)

➢ Radiated (electromagnetic field)

Coupling paths often utilize a complex combination of these methods making the path difficult to identify even when the source and receptor are known. There may be multiple coupling paths and steps taken to attenuate one path may enhance another.

Words and Phrases

electromagnetic interference (EMI) 电磁干扰
electromagnetic compatibility (EMC) 电磁兼容
source of the interference 干扰源
ultimate goal 最终目标
function *v.* 运行
interfere with 干扰，干涉
power line 输电线
out of control 失控
emission *n.* 发射，辐射
FM *abbr.* 调频
CB radio (Citizen's Band radio) 民用波段收音机
stereo *n.* 立体声系统
radio tower 无线电接收塔
airport radar 机场雷达
laptop computer 膝上型计算机
helicopter *n.* 直升机
VCR *abbr.* 录像机
rudder *n.* 舵，方向舵
receptor *n.* 受体，接收器
coupling path 耦合路径
eliminate *v.* 消除
nuclear power plant 核电站
turbine control valve 涡轮控制阀
malfunction *n.* 故障
walkie talkies 步话机
shield *v.* 屏蔽
filter *v.* 滤波
reroute *v.* 改变路线
attenuate *v.* 消弱
unintentional *adj.* 无意识的，无心的
radiate *v.* 辐射
take steps 设法，采取措施
inductively coupled 电感性耦合
capacitively coupled 电容性耦合

Notes

1. Each of these three elements must be present although they may not be readily identified in every situation. Electromagnetic compatibility problems are generally solved by identifying at least two of these elements and eliminating (or attenuating) one of them.

三个要素都一定存在，而在有些情况下它们不容易被识别出来，解决电磁兼容问题一般需要至少识别出两个要素并消除或消弱其中之一。

2. A more thorough and perhaps more secure approach would be to identify the coupling path and take steps to eliminate it.

更彻底而且可能更安全的方法是确定耦合路径并逐步消除之。

3. Potential sources of electromagnetic compatibility problems include radio transmitters, power lines, electronic circuits, lightning, lamp dimmers, electric motors, arc welders and just about anything that utilizes or creates electromagnetic energy.

电磁兼容问题的潜在的干扰源包括无线电发射器、输电线、电子电路、闪电、调光灯、电机、电弧焊接和任何利用或产生电磁能量的东西。

Exercises

1. Answer the following questions according to the text

(1) What is electromagnetic interference (EMI)?

(2) What is electromagnetic compatibility (EMC)?

(3) What are the three essential elements to any EMC problem?

(4) There are four main categories of coupling electromagnetic energy from a source to a receptor. Right or wrong?

(5) Coupling paths often utilize a complex combination of several methods making the path difficult to identify. Right or wrong?

2. Translate the following sentences into Chinese according to the text

(1) Electromagnetic interference (EMI) is a major problem in modern electronic circuits. To overcome the interference, the designer has to either remove the source of the interference, or protect the circuit being affected.

(2) There must be a source of an electromagnetic phenomenon, a receptor (or victim) that cannot function properly due to the electromagnetic phenomenon, and a path between them that allows the source to interfere with the receptor.

(3) The turbine control valves were malfunctioning. The source and the coupling path were originally unknown, however an investigation revealed that the walkie talkies used by the plant employees were the source.

(4) Shielding, filtering, or rerouting the cable, and filtering or redesigning the circuit would then be possible methods of attenuating the coupling path to the point where the problem is non-existent.

(5) In other words, the solution was to eliminate the receptor by making the speed control circuit immune to the electromagnetic phenomenon produced by the source.

3. Translate the following paragraph into Chinese

Coupling can also occur in circuits that share common impedances. For instance, two circuits that share the conductor carrying the supply voltage and the conductor carrying the return path to ground. If one circuit creates a sudden demand in current, the other circuit's voltage supply will drop due to the common impedance both circuits share between the supply lines and the source impedance. This coupling effect can be reduced by decreasing the common impedance. Coupling also can occur with radiated electric and magnetic fields which are common to all electrical circuits. Whenever current changes, electromagnetic waves are generated. These waves can couple over to nearby conductors and interfere with other signals within the circuit.

Text B Designing for Board Level Electromagnetic Compatibility

Although the circuit may be working at the board level, but it may be radiating noise to other parts of the system, causing problems at the system level. So, achieving board level EMC may not be enough. Furthermore, EMC at the system or equipment level may have to satisfy certain emission standards, so that the equipment does not affect other equipment or appliances. Many developed countries have strict EMC standards on electrical equipment and appliances; to meet these, the designer will have to think about EMI suppression—starting from the board level.

In addition to component selection and circuit design, good printed circuit board (PCB) layout is an important factor in EMC performance. Since the PCB is an inherent part of the system, EMC enhancements by PCB layout does not add extra cost towards the finish product. One point to note is that there are no fast and strict rules for PCB layout. There is no single rule that covers all PCB layouts. Most PCB layouts are restricted by board size and the number of copper layers. Some layout techniques may apply to one type of circuit but not another. Much of it will depend on the experience of the PCB layout engineer.

Nevertheless, there are some general rules. These should be treated as general guidelines. One must remember that poor PCB layouts can cause more EMC problems than it can cure, and in many cases, adding filters and components cannot solve the problem. In the end, it may be better to do a complete re-layout of the board. Therefore, good PCB layout practice at the outset is the best cost saving method.

The following are some general guidelines for PCB layout:

- Increase the separation between tracks to minimize crosstalk by capacitive coupling.
- Maximize the PCB capacitance by placing the power and ground in parallel.
- Place sensitive and high frequency tracks far away from high noise power tracks.
- Widen ground and power tracks to reduce the impedance of both power and ground lines.

1. Segmentation

Segmentation is the use of physical separation to reduce the coupling between different types of circuit, particularly by the power and ground tracks.

Fig. 2. 14 shows a typical example of separating four different circuits using the segmentation technique. In the ground plane, the un-metallized moat is used to isolate the four ground planes. Coupling between different circuit power planes is reduced. High speed digital circuits need to be placed near the power supply inlet because of their higher transient power demand. For the L and C, it is better to use different values of L and C filter components instead of one large L and C because they can provide different filtering characteristics for different circuits.

2. Decouple Local Supplies and ICs

Localized decoupling can reduce noise propagating along the supply rail. The use of large bypass capacitors connected at supply entry to the PCB will help as a low frequency ripple filter and potential reservoir for sudden power demands. In addition, decoupling capacitors should be connected between power and ground at each IC, as close as possible to the pins. This helps to filter out switching noises from the IC.

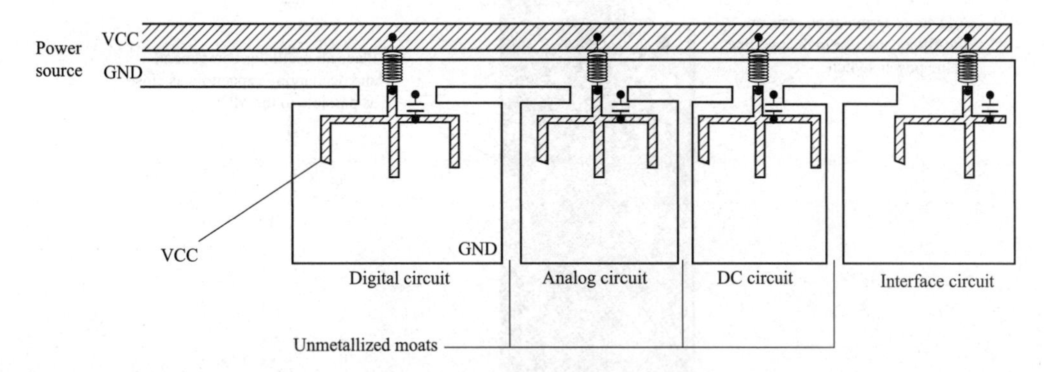

Fig. 2. 14 Separated function blocks

3. RF Current in Reference Plane

Whether with the reference ground plane on multi-layer PCBs or ground traces on single-layer PCBs, a current path is present from the load back to the power supply source. The lower the impedance of the return path has, the better the EMC performance of the PCB has. Long return paths can create mutual coupling because of the RF current from load to source. Therefore, the return path should be as short as possible, the loop area as small as possible.

4. Grounding Techniques

Grounding techniques apply to both multi-layer and single-layer PCBs. The objective of grounding techniques is to minimize the ground impedance and thus to reduce the potential of the ground loop from circuit back to the supply.

On a single-layer (single sided) PCB, the width of the ground track should be as wide as possible, at least 1. 5 mm (60mil). Since the star arrangement is impossible on single-layer PCBs, the use of jumpers and changes in ground track width should be kept to a minimum, as these cause change in track impedance and inductance.

On a double-layer (double sided) PCB, the ground grid/matrix arrangement is preferred for digital circuits because this arrangement can reduce ground impedance, ground loops, and signal return loops. As with the single-layer PCBs, the width of the ground and power tracks should be at least 1. 5 mm. Another scheme is to have a ground plane on one side, the signal and power line on the other side. In this arrangement the ground return path and impedance will be further reduced and decoupling capacitors can be placed as close as possible between the IC supply line and the ground plane.

Fig. 2. 15 shows some improvements on a typical printed circuit board used in a washing machine. Fig. 2. 16 shows some improvements on a typical printed circuit board used in an air conditioner.

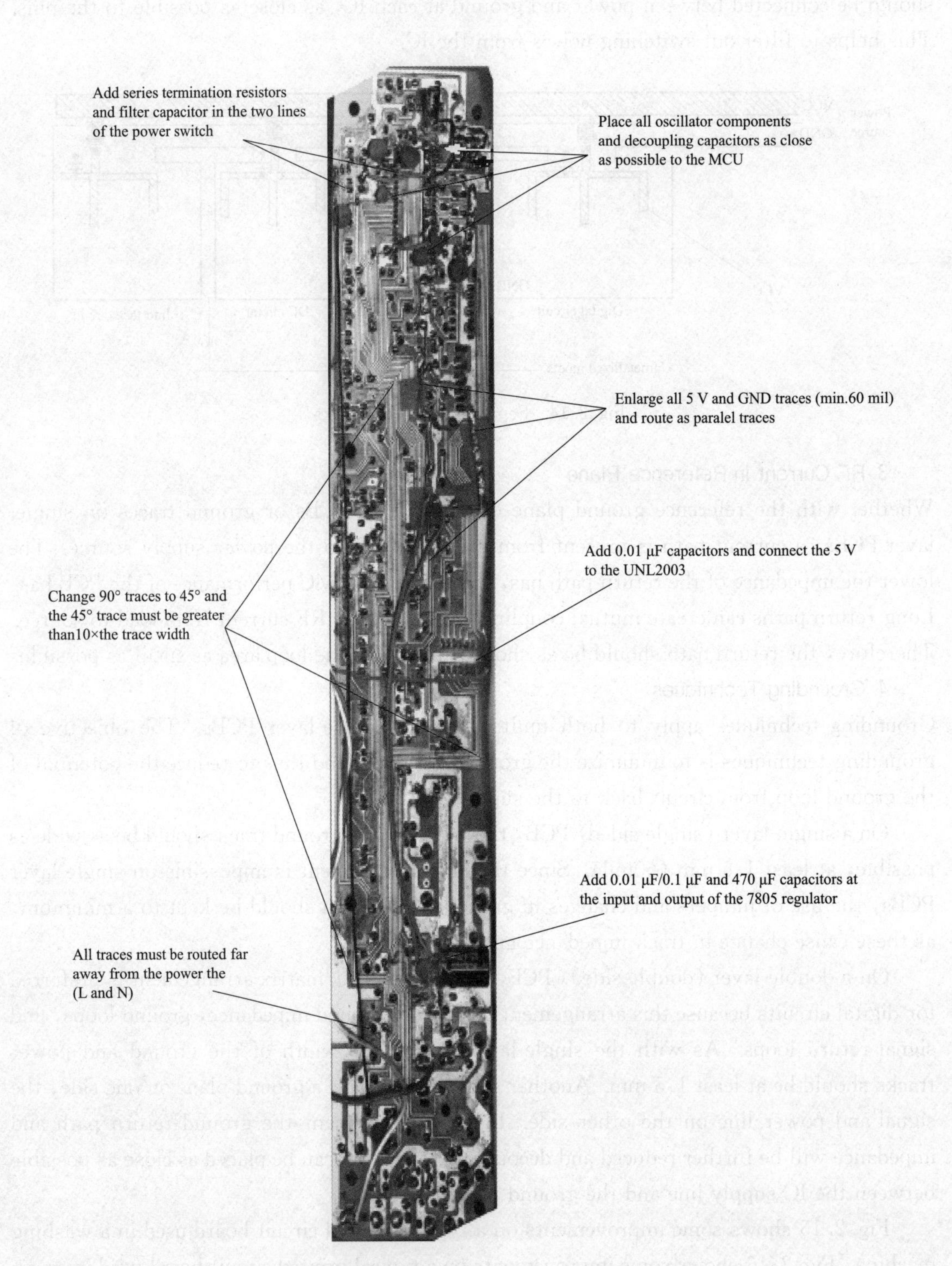

Fig. 2. 15 PCB Improvement—Example 1

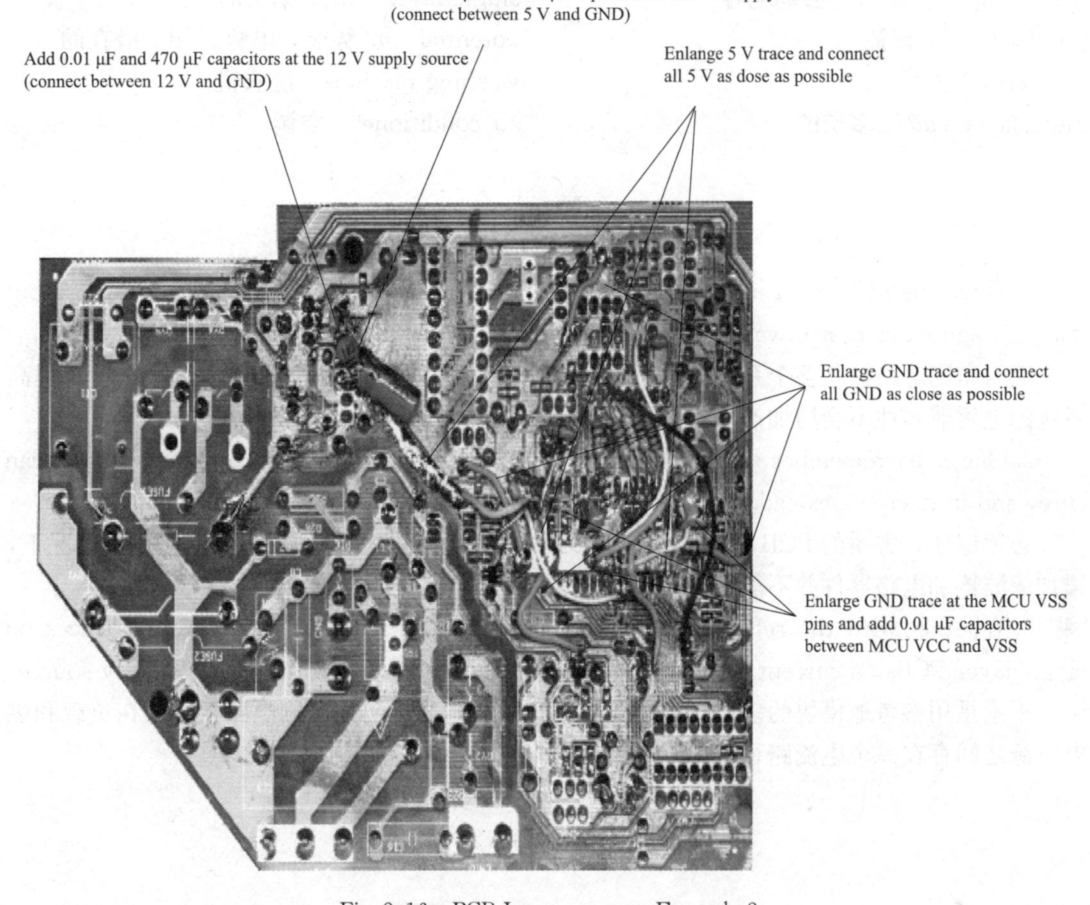

Fig. 2.16 PCB Improvement—Example 2

Words and Phrases

board *n.* （电路）板
level *n.* 标准，级别
equipment *n.* 设备，装置
printed circuit board (PCB) 印刷电路板
layout *n.* 设计，规划
filter *n.* 滤波器
separation *n.* 分离，分开
improvement *n.* 改进
component selection 器件的选择
circuit design 电路的设计
capacitive coupling 容性耦合
ground track 地线
power track 电源线
track *n.* 轨迹，路，线
crosstalk *n.* 交叉，串话
impedance *n.* 阻抗
segmentation *n.* 分割
plane *n.* 平面，模块
moat *n.* 护城河，壕沟
decoupling *n.* 去耦，解耦（装置）
localized decoupling 局部解耦
inlet *n.* 入口
IC *abbr.* 集成电路
interface circuit 接口电路
bypass capacitor 旁路电容
ripple filter 纹波滤波器

decoupling capacitor　去耦电容
propagate　*v.* 传播
RF　*abbr.* 射频
multi-layer　*adj.* 多层的
single-layer　*adj.* 单层的
potential　*n.* 势能，电势；adj. 潜在的
washing machine　洗衣机
air conditioner　空调

Notes

1. Since the PCB is an inherent part of the system, EMC enhancements by PCB layout does not add extra cost towards the finish product.

因为印刷电路板是整个系统必不可少的一部分，所以通过很好地印刷电路板设计来提高系统的电磁兼容性不会增加产品的额外成本。

2. One must remember that poor PCB layouts can cause more EMC problems than it can cure, and in many cases, adding filters and components cannot solve the problem.

必须记住，劣质的 PCB 设计能引起更多的电磁兼容问题而且很难解决，在很多情况下，增加滤波器和电路器件并不能解决这些问题。

3. Whether with the reference ground plane on multi-layer PCBs or ground traces on single-layer PCBs, a current path is present from the load back to the power supply source.

不论是用参考地模块的多层印刷电路板，还是用地线地单层印刷电路板，都在负载和供电电源之间存在一个电流路径。

Exercises

1. Answer the following questions according to the text

(1) Achieving board level EMC is enough for a system. Right or wrong?

(2) Does EMC enhancements by PCB layout add extra cost towards the finish product?

(3) Good PCB layout practice at the outset is the best cost saving method. Right or wrong?

(4) Do you know some general guidelines for PCB layout?

(5) Decoupling capacitors should be connected between power and ground at each IC, as close as possible to the pins. This helps to filter out switching noises from the IC. Right or wrong?

2. Translate the following sentences into Chinese according to the text

(1) Furthermore, EMC at the system or equipment level may have to satisfy certain emission standards, so that the equipment does not affect other equipment or appliances.

(2) One point to note is that there are no fast and strict rules for PCB layout. There is no single rule that covers all PCB layouts. Most PCB layouts are restricted by board size and the number of copper layers.

(3) High speed digital circuits need to be placed near the power supply inlet because of their higher transient power demand.

(4) Long return paths can create mutual coupling because of the RF current from load to source. Therefore, the return path should be as short as possible, the loop area as small as possible.

(5) As with the single-layer PCBs, the width of the ground and power tracks should be at least 1.5 mm. Another scheme is to have a ground plane on one side, the signal and power line on the other side.

3. Translate the following paragraph into Chinese

The majority of modern digital integrated circuits (IC) are manufactured using complementary metal oxide semiconductor (CMOS) technology. The static power consumption of CMOS devices may be lower, but with fast switching rates the CMOS device demands transient power from the supply. The dynamic power demand of a high speed clocked CMOS device may exceed an equivalent bipolar device. Therefore, decoupling capacitors must be used on these devices to reduce the transient power demand from the power supply.

Unit 5 Electrotechnical Materials and the Characteristics

Text A Introduction of the Electrotechnical Materials

Soft magnetic materials are used in devices that are subjected to alternating magnetic fields and in which energy losses must be low; one familiar example consists of transformer cores. For this reason the relative area within the hysteresis loop must be small; it is characteristically thin and narrow. Consequently, a soft magnetic material must have a high initial permeability and a low coercivity. A material possessing these properties may reach its saturation magnetization with a relatively low applied field and still has low hysteresis energy losses. The saturation field or magnetization is determined only by the composition of the material. For example, in cubic ferrites, substitution of a divalent metal ion such as for in will change the saturation magnetization. However, susceptibility and coercivity which also influence the shape of the hysteresis curve, are sensitive to structural variables rather than to composition. For example, a low value of coercivity corresponds to the easy movement of domain walls as the magnetic field changes magnitude and/or direction. Structural defects such as particles of a nonmagnetic phase or voids in the magnetic material tend to restrict the motion of domain walls, and thus increase the coercivity. Consequently, a soft magnetic material must be free of such structural defects. Another property consideration for soft magnetic materials is electrical resistivity. In addition to the hysteresis energy losses described above, energy losses may result from electrical currents that are induced in a magnetic material by a magnetic field that varies in magnitude and direction with time; these are called eddy currents. It is most desirable to minimize these energy losses in soft magnetic materials by increasing the electrical resistivity. This is accomplished in ferromagnetic materials by forming solid solution alloys; iron-silicon and iron-nickel alloys are examples. The ceramic ferrites are commonly used for applications requiring soft magnetic materials because they are intrinsically electrical insulators. In addition, soft magnetic materials are used in generators, motors, dynamos, and switching circuits.

1. Hard Magnetic Materials

Hard magnetic materials are utilized in permanent magnets, which must have a high resistance to demagnetization. In terms of hysteresis behavior, a hard magnetic material has a high remanence, coercivity, and saturation flux density, as well as a low initial permeability, and high hysteresis energy losses. The two most important characteristics relative to applications for these materials are the coercivity and what is termed the "energy product", designated as BH_{max}. This BH_{max} corresponds to the area of the largest B-H rectangle that can be constructed within the second quadrant of the hysteresis curve. The value of the energy product is representative of the energy required to demagnetize a permanent magnet; that is, the larger BH_{max} the harder is the material in terms of its magnetic characteristics. Again, hysteresis behavior is related to the ease with which the magnetic domain boundaries move; by impeding domain wall motion, the coercivity and susceptibility are enhanced, such that a large external field is required for demagnetization. Furthermore, these characteristics are interrelated to the microstructure of the material.

2. Dielectric Materials

A number of ceramics and polymers are utilized as insulators and/or in capacitors. Many of the ceramics, including glass, porcelain, steatite, and mica, have dielectric constants within the range of 6 to 10. These materials also exhibit a high degree of dimensional stability and mechanical strength. Typical applications include power line and electrical insulation, switch bases, and light receptacles. The titania (TiO_2) and titanate ceramics, such as barium titanate ($BaTiO_3$), can be made to have extremely high dielectric constants, which render them especially useful for some capacitor applications. The magnitude of the dielectric constant for most polymers is less than for ceramics, since the latter may exhibit greater dipole moments: values for polymers generally lie between 2 and 5. These materials are commonly utilized for insulation of wires, cables, motors, generators, and so on, and, in addition, for some capacitors.

3. Conductive Materials

Most metals are extremely good conductors of electricity because of the large numbers of free electrons that have been excited into empty states above the Fermi energy. At this point it is convenient to discuss conduction in metals in terms of the resistivity, the reciprocal of conductivity; the reason for this switch in topic should become apparent in the ensuing discussion. Since crystalline defects serve as scattering centers for conduction electrons in metals, increasing their number raises the resistivity (or lowers the conductivity). The concentration of these imperfections depends on temperature, composition, and the degree of cold work of a metal specimen. In fact, it has been observed experimentally that the total resistivity of a metal is the sum of the contributions from thermal vibrations, impurities, and plastic deformation; that is, the scattering mechanisms act independently of one another. This may be represented in mathematical form as follows:

$$\rho_{total} = \rho_t + \rho_i + \rho_d \tag{2-4}$$

in which ρ_t, ρ_i and ρ_d represent the individual thermal, impurity, and deformation resistivity contributions, respectively. Equation (2-4) is sometimes known as Matthiessen's rule. Electrical and other properties of copper render it the most widely used metallic conductor. Oxygen-free high-conductivity (OFHC) copper, having extremely low oxygen and other impurity contents, is produced for many electrical applications. Aluminum, having a conductivity only about one-half that of copper, is also frequently used as an electrical conductor. Silver has a higher conductivity than either copper or aluminum; however, its use is restricted on the basis of cost. For some applications, such as furnace heating elements, a high electrical resistivity is desirable. The energy loss by electrons that are scattered is dissipated as heat energy. Such materials must have not only a high resistivity, but also a resistance to oxidation at elevated temperatures and, of course, a high melting temperature. Nichrome, a nickel-chromium alloy, is commonly employed in heating elements.

4. Piezoelectric Materials

An unusual property exhibited by a few ceramic materials is piezoelectricity, or, literally, pressure electricity: polarization is induced and an electric field is established across a specimen by the application of external forces. Reversing the sign of an external force (i. e., from tension to compression) reverses the direction of the field. Piezoelectric materials are utilized in transducers, which are devices that convert electrical energy into mechanical strains, or vice versa. Some other familiar applications that employ piezoelectrics include phonograph cartridges, microphones, speakers, audible alarms, and ultrasonic imaging. Piezoelectric materials include titanates of barium and lead, lead zirconate ammonium dihydrogen phosphate and quartz. This property is characteristic of materials having complicated crystal structures with a low degree of symmetry. The piezoelectric behavior of a polycrystalline specimen may be improved by heating above its Curie temperature and then cooling to room temperature in a strong electric field.

Words and Phrase

divalent *adj*. 化合物二价的
magnetic domain 磁畴
dielectric *n*. 电解质，绝缘体；
adj. 非传导性的
polymer *n*. 聚合体，多聚体
receptacles *n*. 插座，容器，花托
titanate *n*. 钛酸盐
barium *n*. 钡
piezoelectric *adj*. 压电的
tension *n*. 张力
compression *n*. 压缩
strain *n*. 应力
zirconate *n*. 锆酸盐
ammonium *n*. 铵
dihydrogen *adj*. 二氢的
phosphate *n*. 磷酸盐
Curie temperature 居里温度
crystalline *adj*. 结晶的
reciprocal *adj*. 互惠的，倒数的；*n*. 倒数
furnace *n*. 火炉，熔炉
scatter *v*. 撒播，散开，四散，使分散，驱散
nichrome *n*. 镍铬合金
chromium *n*. 铬

Notes

1. For this reason the relative area within the hysteresis loop must be small; it is characteristically thin and narrow.

因此，磁滞回线内的相对面积必须很小，其特点是狭而窄。

2. In terms of hysteresis behavior, a hard magnetic material has a high remanence, coercivity, and saturation flux density, as well as a low initial permeability, and high

hysteresis energy losses.

在磁滞特性方面，永磁材料具有高的剩磁、矫顽力和饱和磁通密度，以及低的初始磁导率和高的磁滞能耗。

3. These materials also exhibit a high degree of dimensional stability and mechanical strength. Typical applications include power line and electrical insulation, switch bases, and light receptacles.

这些材料还表现出高度的形稳性和机械强度，典型应用包括电源线和电气绝缘、开关底座和灯插座。

4. Such materials must have not only a high resistivity, but also a resistance to oxidation at elevated temperatures and, of course, a high melting temperature.

这种材料不仅具有高电阻率，而且在高温下也具有抗氧化性，当然也具有高熔点温度。

Exercises

1. Answer the following questions according to the text

(1) What is determined only by the composition of the material?

(2) What are the main characteristics of the hard magnetic material?

(3) Silver has a lower conductivity than either copper or aluminum. Right or wrong?

(4) What is the function of Piezoelectric materials when they are utilized in transducers?

2. Translate the following sentences into Chinese according to the text

(1) Consequently, a soft magnetic material must have a high initial permeability and a low coercivity.

(2) Again, hysteresis behavior is related to the ease with which the magnetic domain boundaries move; by impeding domain wall motion, the coercivity and susceptibility are enhanced.

(3) The concentration of these imperfections depends on temperature, composition, and the degree of cold work of a metal specimen.

(4) The piezoelectric behavior of a polycrystalline specimen may be improved by heating above its curie temperature and then cooling to room temperature in a strong electric field.

3. Translate the following paragraph into Chinese

Giant Magnetostrictive Material (GMM) is a new functional material which has bi-directional transduction effects, it can be made into Giant Magnetostrictive Actuator (GMA) by using the direct effect, also can be made into sensors by using the converse effect. Compared with piezoelectric actuator, GMA owns the advantages of larger load capacity, rapider response, higher reliability and lower voltage drive, etc. It has a promising prospect in precision positioning, active noise and vibration control and fluid control (pump, valve) fields.

Text B Applications of Magnetostrictive Material

Terfenol-D is said to produce "giant" magnetostriction, strain greater than any other commercially available smart material. Magnetic domains in the crystal rotate when a magnetic field is applied, which providing proportional, positive and repeatable expansion in microseconds.

The mechanism of magnetostriction can act as a very sensitive sensor by converting mechanical work into electrical energy. Rather than having an external magnetic field force the rotation of magnetic domains to deform the crystal to do mechanical work, devices can be made where mechanical vibrations or loads deform the crystal and, in doing so, rotate the magnetic domains, which generates a transient current. While most attention has been focused on the active aspect of the material (electrical-to-mechanical), several different applications have recently begun to exploit the reverse phenomenon of the material, commonly called to Villari Effect.

Magnetostrictive technology is found in numerous applications throughout many different industries. Samples of the types of problems that have been solved with Terfenol-D include:

- Active noise and vibration cancellation
- Sonar
- Fuel injection
- Medical
- Sonochemistry
- Nozzle anti-clogging system
- Screening applications
- Metals casting industry
- Fast tool servo

Active Noise and Vibration Cancellation (ANVC)

In general, active noise and vibration cancellation works by injecting an equal but opposite signal into a system in order to counteract an undesirable noise and/or vibration. Therefore, an understanding of the magnitude and phase relationships in the system is critical in this endeavor. Although every situation is unique, the overall approach in implementing an ANVC system is similar from one application to the next. The basic approach includes the following high-level tasks:

- System Identification
- Actuator and Sensor Placement
- System Specifications Development
- Actuator Design
- Controller Algorithm Development and Implementation

➢ System Testing and Evaluation

The first task addressed in finding a solution to the gear mesh noise problem was system identification. This involved a modal analysis to correlate the input vibrations from the gear housing to the vibrations along the axle of the vehicle. This analysis provided transfer function information so that the output at a given location could be predicted from a known input.

The propagation path of the noise was identified next in order to correlate the noise in the cabin of the vehicle with the vibrations along the axle. For this task, engineers need to analyze acceleration and acoustic data. This information was used to determine the most effective placement of actuators and sensors along the axle in order to affect the greatest reduction in gear mesh noise as heard in the vehicle cabin.

The requirements generated in the previous step dictated the specifications of the actuator to be used to counteract the undesirable vibrations in the axle. In some cases, one or more of actuators may be used in the application if the specifications already meet the requirements. Otherwise, a custom actuator may need to be developed. In particular, if there are unique environmental issues that must be addressed, some modifications may be required. Fig. 2. 17 shows the giant magnetostrictive Actuator used in an ANVC system.

The next step in solving a noise and/or vibration problem with active means is to identify a control algorithm that suits the requirements of the system. There are many algorithms to choose from and many are well documented in the industry literature. In the vehicle application, an adaptive feed-forward algorithm that utilized the shaft rotation as a reference signal was incorporated. A Least Mean Square (LMS) algorithm was also used to minimize an "error" signal, where the error signal was the acceleration at a given point along the axle that had a strong correlation to the cabin noise.

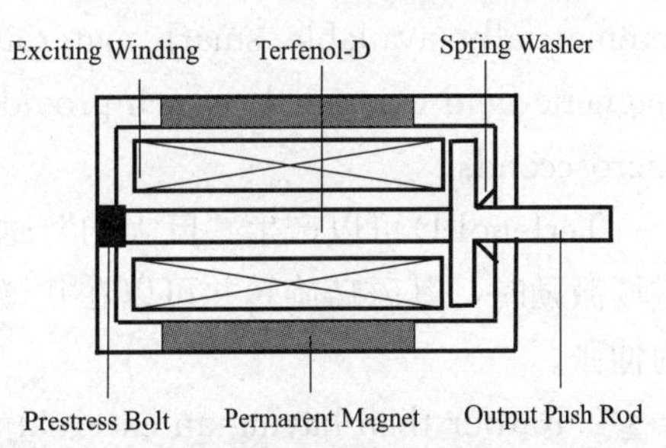

Fig. 2. 17 Giant magnetostrictive actuator

The final step in the ANVC solution is the implementation of the system. This includes subsystem testing as well as full integration. Small adjustments to the controller coefficients are usually necessary to account for unforeseen interactions between system components and any differences between the theoretical design and the actual hardware built.

Words and Phrases

giant magnetostriction 巨磁致伸缩
crystal *n.* 晶体
rotate *v.* 旋转
expansion *n.* 扩张，伸长
microsecond *n.* 微秒
exploit *v.* 开发
Villari Effect 维拉里效应
phonochemistry *n.* 声化学

active noise and vibration cancellation 主动去噪和消振
nozzle anti-clogging system 喷嘴抗堵塞系统
metals casting industry 金属铸造工业
fast tool servo 快速加工伺服系统
counteract *v.* 抵消，中和
identification *n.* 辨识
placement *n.* 布置，放置
system testing and evaluation 系统检测和评估
gear mesh 齿轮啮合
vehicle *n.* 车辆
axle *n.* 车轴，轮轴
transfer function 传递函数
acceleration *n.* 加速度
propagation path 传播路径
acoustic *adj.* 声学的
adaptive feed-forward algorithm 自适应前馈算法
reference signal 参考（基准）信号
Least Mean Square (LMS) algorithm 最小均方差算法
subsystem *n.* 子系统
integration *n.* 综合
adjustment *n.* 调整，调节器
coefficient *n.* 系数
unforeseen *adj.* 无法预测的

Notes

1. Terfenol-D is said to produce "giant" magnetostriction，strain greater than any other commercially available smart material. Magnetic domains in the crystal rotate when a magnetic field is applied，which providing proportional，positive and repeatable expansion in microseconds.

Terfenol-D 可以产生“巨大的”磁致伸缩，其应变远大于其他商业智能材料。当它受到磁场激励时，其磁畴旋转并可以在几微秒的时间内产生与激励磁场呈正比的、正的、可重复的伸张。

2. Rather than having an external magnetic field force the rotation of magnetic domains to deform the crystal to do mechanical work，devices can be made where mechanical vibrations or loads deform the crystal and，in doing so，rotate the magnetic domains，which generates a transient current.

外加磁场使磁畴旋转导致晶体变形产生机械能，与此相当，机械振动或加载导致晶体变形及磁畴旋转可以产生瞬时电流。

Exercises

1. Answer the following questions according to the text

(1) Do you know some applications of magnetostrictive material? What are they?

(2) Although every situation is unique，the overall approach in implementing an ANVC system is similar from one application to the next. Right or wrong?

(3) The propagation path of the noise was identified next in order to correlate the noise in the cabin of the vehicle with the vibrations along the axle. For this task，engineers need to analyze acceleration and acoustic data. Right or wrong?

(4) The final step in the ANVC solution is the implementation of the system. Right or wrong?

2. Translate the following sentences into Chinese according to the text

(1) The mechanism of magnetostriction can act as a very sensitive sensor by converting mechanical work into electrical energy.

(2) In general, active noise and vibration cancellation works by injecting an equal but opposite signal into a system in order to counteract an undesirable noise and/or vibration.

(3) Therefore, an understanding of the magnitude and phase relationships in the system is critical in this endeavor.

(4) The first task addressed in finding a solution to the gear mesh noise problem was system identification.

(5) Small adjustments to the controller coefficients are usually necessary to account for unforeseen interactions between system components and any differences between the theoretical design and the actual hardware built.

3. Translate the following paragraph into Chinese

Magnetostrictive (MS) linear position sensors constitute an interesting possibility as generic all-use sensors because they are by principle non-contact and absolute. These are very desirable characteristics for machine tool operation, as well as their non-optical nature which makes them resistant to typical contaminants of the machine tool environment like shavings and metalworking coolant fluid.

Part Ⅲ Power Electronics and Power Drives

Unit 1 Emerging System Applications and Technological Trends in Power Electronics

Text A Emerging System Applications of Power Electronics

Power electronics (PE) is going through an exciting time, with increasingly widespread applications. PE fits a typical definition of high technology, since it draws expertise from three broad technical fields of electrical engineering: circuits and automatic control, magnetic and electric machines, and semiconductor devices and integrated circuits (IC). Today, PE has outgrown its classic definition. It has evolved to include just about all aspects of electrical and electronic engineering. It includes analog and digital circuits, converter circuits, magnetic and electric machines, linear and nonlinear control, energy generation and storage, system engineering and integration, radiofrequency (RF) circuits, antennas, ICs and monolithic passives, and power semiconductors and ICs. This text summarizes six emerging system applications of PE, which include green energy system integrations; microgrids; all things grid connected; transportation electrification; smart homes, smart buildings, and smart cities; and energy harvesting.

1. Green Energy System Integration

One of the most significant system applications of PE is the integration of green energy systems. Solar and wind energy systems are maturing fast to penetrate deeply into the energy market to provide electricity for average consumers. The technologies for individual systems are mature, and recent trends are toward the integration of solar farms with wind farms, particularly with offshore wind farms. To transmit the power from offshore to onshore, new highvoltage direct current (HVDC) transmission technology has been developed and deployed around the world. The new technology leverages the recent progress in PE, such as HV insulated gate bipolar transistors (IGBT) and multilevel voltage-source converter (VSC) circuit topology.

2. Microgrids

A microgrid is an autonomous electric power subsystem. It can be DC, AC, or a hybrid of DC and AC. A structured microgrid is integrated with loads, energy sources, storage devices, sensors, and data buses, and is an autonomous subsystem. DC microgrids are particularly advantageous for practice due to their many benefits, which include higher efficiency, more robust system operation, no impedance matching issues, no synchronization, simple waveforms, and a large body of knowledge in PE and systems being directly leveraged.

3. All Things Grid Connected

"All things grid connected" is a phrase used to describe the key hardware types that an electronictized grid may encompass. They are grid-tied inverters that connect renewable energy resources to the grid, digitally controlled solid-state power transformers, smart fuses and solid-state circuit breakers, flexible AC transmission systems, static variable reactive power compensators, multilevel power converters for HVDC, bidirectional multiport power and control units for the integration of multiple resources, and energy storage, to name a few. In other words, PE is playing a crucial role in grid modernization. It can be said that in the competition for the dominance of smart grids, those who possess better PE technologies will have the upper hand, as was illustrated recently by the success of new HVDC technology with VSC, together with the breakthrough in HVDC circuit breakers.

4. Transportation Electrification

Another important area that is dominant in emerging system applications is transportation electrification. PE is the essential driving force of electrification, at least at the early stages. In 2014, the IEEE Transportation Electrification Initiative (TEI) was transitioned into the IEEE Transportation Electrification Community (TEC), with Power Electronics Society (PELS) providing its administrative home. The community covers all aspects of transportation electrification, such as electric vehicles, charging stations, grid-connected chargers, more electric airplanes, more electric ships, off-road vehicles, trains and traction, connected vehicles, self-driving cars, and intelligent highways.

5. Smart Homes, Smart Buildings, and Smart Cities

A closely related system application, but at a higher level to transportation electrification and sustainable energies, is the initiative for smart homes, smart buildings, and smart cities. Smart homes are a natural growth from the smart grid, since the power will be from the smart grid, together with local renewable generation. The scope of smart homes is broad and ever evolving, but it generally includes the following features: smart appliances, smart lighting, personalized energy, mobility, mobile communication, smart security, and remote personal care. To carry out smart control and monitoring, a smart (remote) sensor network is needed. For smart buildings, the key features include integration of renewable energies, high-voltage ac or medium-voltage DC control and distribution, lighting control, access control, security, and power and cooling control.

The idea of smart cities is very recent but nevertheless can be considered a natural extension of smart homes and smart buildings. The IEEE has a workforce called the IEEE Smart Cities Initiative, of which PELS is a member. The scope of a smart city is even more broad and ever evolving, but its key features include smart water, smart energy, smart buildings, smart waste, smart public services, smart security, and smart mobility. PE will play an important role in providing the backbone for future energy conversion and control intelligence for future smart cities.

6. Energy Harvesting

Energy harvesting is gaining much wider recognition lately because of the proliferation of mobile technologies and low-power microelectronics. There are many different forms of energy harvesting. Most prevalent are RF, vibration, thermal, solar, human bodies, and structural energy harvesting. An energy-harvesting (mini) system typically includes four parts: energy-conversion device (s), harvesting, power conditioning, and storage. All these fields are closely related to PE.

Words and Phases

power electronics 电力电子
semiconductor device 半导体器件
integrated circuit (IC) 集成电路 (IC)
radio frequency (RF) 射频 (RF)
green energy 绿色能源
microgrid *n.* 微电网
transportation electrification 交通电气化
smart *adj*. 智能的
energy harvesting 能量收集
solar/wind farm 太阳能发电厂、风力发电厂
highvoltage direct current (HVDC) transmission 高压直流 (HVDC) 输电
insulated gate bipolar transistors (IGBT) 绝缘栅双极型晶体管 (IGBT)
voltage-source converter (VSC) 电压源型变流器 (VSC)
autonomous *adj*. 自治的
robust *adj*. 鲁棒
synchronization *n.* 同步
solid-state power transformer 固态变压器
solid-state circuit breaker 固态断路器
flexible AC transmission 柔性交流输电
multilevel power converter 多电平功率变换器
electric vehicle 电动汽车
self-driving car 自动驾驶汽车
sensor *n.* 传感器
distribution *n.* 分配
low-power microelectronic 低功耗微电子
vibration *n.* 振动
thermal *adj*. 热的

Notes

1. IEEE

电气和电子工程师协会（IEEE，全称是 Institute of Electrical and Electronics Engineers）是一个美国的电子技术与信息科学工程师的协会，是目前世界上最大的非营利性专业技术学会，其会员人数超过 40 万人，遍布 160 多个国家。IEEE 致力于电气、电子、计算机工程和与科学有关的领域的开发和研究，在太空、计算机、电信、生物医学、电力及消费性电子产品等领域已制定了 900 多个行业标准，现已发展成为具有较大影响力的国际学术组织。

2. Today, PE has outgrown its classic definition. It has evolved to include just about all aspects of electrical and electronic engineering. It includes analog and digital circuits, converter circuits, magnetic and electric machines, linear and nonlinear control, energy generation and storage, system engineering and integration, radiofrequency (RF) circuits,

antennas, ICs and monolithic passives, and power semiconductors and ICs.

现今，PE 已经超越了它的经典定义。它已经发展到包括电气和电子工程的所有方面，包括模拟和数字电路，变换器电路，磁和电机，线性和非线性控制，能量产生和存储，系统工程和集成，射频电路，天线，IC 和单片无源器件，以及功率半导体和 IC。

Exercises

1. Answer the following questions according to the text

(1) Why does PE fit a typical definition of high technology?

(2) What are the characteristics of DC microgrids?

(3) Please list some aspects of transportation electrification.

(4) What are the key features of the smart city?

(5) What are the forms of energy harvesting?

2. Translate the following sentences into Chinese according to the text

(1) Solar and wind energy systems are maturing fast to penetrate deeply into the energy market to provide electricity for average consumers.

(2) DC microgrids are particularly advantageous for practice due to their many benefits, which include higher efficiency, more robust system operation, no impedance matching issues, no synchronization, simple waveforms, and a large body of knowledge in PE and systems being directly leveraged.

(3) It can be said that in the competition for the dominance of smart grids, those who possess better PE technologies will have the upper hand, as was illustrated recently by the success of new HVDC technology with VSC, together with the breakthrough in HVDC circuit breakers.

(4) PE will play an important role in providing the backbone for future energy conversion and control intelligence for future smart cities.

3. Translate the following paragraph into Chinese

Energy harvesting is gaining much wider recognition lately because of the proliferation of mobile technologies and low-power microelectronics. There are many different forms of energy harvesting. Most prevalent are RF, vibration, thermal, solar, human bodies, and structural energy harvesting. An energy-harvesting (mini) system typically includes four parts: energy-conversion device (s), harvesting, power conditioning, and storage. All these fields are closely related to PE.

Text B Technological Trends in Power Electronics

The four major technology trends playing a crucial role in driving the emerging system applications include adiabatic power conversion, monolithic power conversion, multilevel power converters, and wide-bandgap devices.

1. Power Conversion Goes Adiabatic

With the proliferation of PE applications in just about every aspect of human life, the requirement for energy efficiency is becoming ever more stringent. For instance, for a hand-held device, be it a cell phone or an iPad, long battery life is essential. Also, to pack increasingly more functions (hardware) into a small space, the thermal energy dissipated by electronics has to be minimized. Both power processing and management circuitry must be adiabatic, meaning that they generate virtually no heat. Fig. 3. 1 shows a definition of adiabatic power conversion. As shown, it is the region of the curve where the efficiency for power conversion is ⩾98. 5%. The efficiency is significantly higher across a wide variation of load.

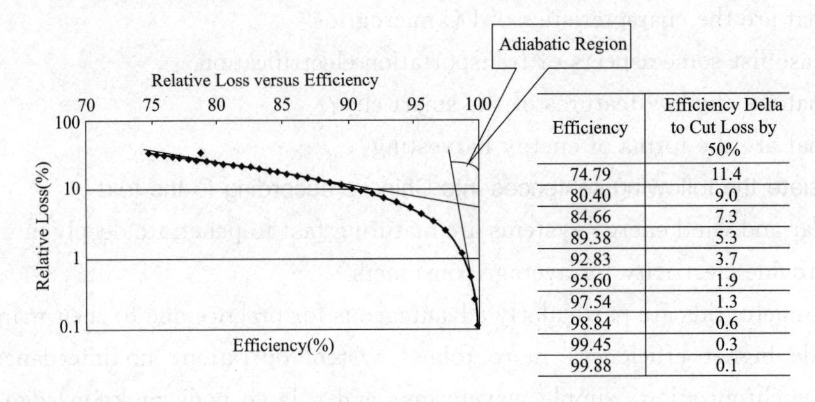

Efficiency	Efficiency Delta to Cut Loss by 50%
74.79	11.4
80.40	9.0
84.66	7.3
89.38	5.3
92.83	3.7
95.60	1.9
97.54	1.3
98.84	0.6
99.45	0.3
99.88	0.1

Fig. 3. 1 Definition of adiabatic power conversion

2. Low-Power Conversion Goes Monolithic

For low-power applications, say 130 W, the power processing technology is going monolithic. Many electronic gadgets can be included in this category. Going monolithic allows the designer to have a high-level integration for more functionality in the next-higher assembly, the power density to be increased for denser package, and production to reduce cost by leveraging the masseconomic scale of silicon foundry production capability.

Significant technical progress has been made in recent years. Perhaps the first product to employ a monolithic switching power converter is by Intel, where distributed multiple power converters were used to power the multicore chips. This distributed power processing allows the power elements to be distributed around the chips they power and operate in a distributed manner, reducing the heat they generate. This is ideal for a multicore architecture. It is anticipated that product offerings of the monolithic integrated power converters will come to market in 2016 or 2017. This success represents a dream come true for PE engineers, that is, to have a monolithic switching converter on an IC.

3. High-Power Conversion Goes Multilevel

Another dream for PE engineers is to have an inverter that can produce a sinusoidal output without much filtering. This dream is also coming true for us in the success of multilevel power converter technology. A multilevel converter allows low-voltage rated parts to be used for high-voltage levels by series stacking them, together with static and dynamic voltage share circuitry. A multilevel converter can also have multiple parallel strings to accommodate for more power (current). The ability to configure the devices in series or in parallel strings for multiphase configuration opens wider applications. In some sense, this scenario parallels that of classic multiphased AC transformers, where a designer can mix and match various different phases to optimize the design for a particular application (requirements). Multilevel power conversion is dominant in high-power applications in industrial drives, grid applications, HVDC, and many others.

4. Wide-Bandgap Devices Are Going Mainstream

Wide-bandgap semiconductor devices are gaining momentum in real-world applications. Serious product offerings are gaining market share in mainstream applications. These are generally in two areas: high power and high speed. For high-power applications, silicon carbide (SiC) -based device and gallium nitride (GaN) -based switching device will gain the upper hand relative to metal oxide semiconductor (MOS) -based devices because of its ability to switch fast and withstand high voltages and high temperatures. SiC-based devices are ahead of the curve in terms of product availability in the market. GaN devices are gaining momentum in high-speed applications such as RF power amplifiers. The challenges for GaN power switching devices are enhancement-mode operation similar to a MOSFET, stable dynamic on-resistance, efficient gate drive circuitry, and efficient controllers. GaN will have to compete for low-to-medium power applications, where lateral double-diffused MOS can still be viable. The silicon [complementary MOS (CMOS)] controller is here for a long time to come because of its low power losses. To summarize, SiC devices will have the edge for high-power application; GaN devices will have an edge in high-speed applications; and CMOS will still have an edge for low-power, high-efficiency applications.

Words and Phases

adiabatic *adj*. 绝热的
monolithic *adj*. 单片的
wide-bandgap device 宽禁带器件
hand-held device 手持设备
pack *v*. 封装
electronic gadget 电子产品
power density 功率密度
silicon *n*. 硅
multicore chip 多核芯片
monolithic integrated power converter 单片集成功率变换器
high power 高功率
multiphased ac transformer 多相交流变压器
industrial drive 工业驱动
mainstream *n*. 主流
silicon carbide (SiC) 碳化硅 (SiC)
gallium nitride (GaN) 氮化镓 (GaN)

Notes

1. The four major technology trends playing a crucial role in driving the emerging system

applications include adiabatic power conversion, monolithic power conversion, multilevel power converters, and wide-bandgap devices.

在推动新兴系统应用中发挥关键作用的四大技术趋势包括绝热功率变换，单片功率变换，多电平功率变换器和宽禁带器件。

2. To summarize, SiC devices will have the edge for high-power application; GaN devices will have an edge in high-speed applications; and CMOS will still have an edge for low-power, high-efficiency applications.

总之，SiC 器件将具有大功率应用的优势；GaN 器件在高速应用中具有优势；而 CMOS 仍将在低功耗、高效率应用中具有优势。

Exercises

1. Answer the following questions according to the text

(1) What are the major technology trends in power electronics?

(2) What is the definition of adiabatic power conversion?

(3) What is the benefit of going monolithic?

(4) Why are the wide-bandgap semiconductor devices gaining momentum in real-world applications?

2. Translate the following sentences into Chinese according to the text

(1) With the proliferation of PE applications in just about every aspect of human life, the requirement for energy efficiency is becoming ever more stringent.

(2) Going monolithic allows the designer to have a high-level integration for more functionality in the next-higher assembly, the power density to be increased for denser package, and production to reduce cost by leveraging the mass economic scale of silicon foundry production capability.

(3) A multilevel converter allows low-voltage rated parts to be used for high-voltage levels by series stacking them, together with static and dynamic voltage share circuitry.

3. Translate the following paragraph into Chinese

Wide-bandgap semiconductor devices are gaining momentum in real-world applications. Serious product offerings are gaining market share in mainstream applications. These are generally in two areas: high power and high speed. For high-power applications, silicon carbide (SiC) -based device and gallium nitride (GaN) -based switching device will gain the upper hand relative to metal oxide semiconductor (MOS) -based devices because of its ability to switch fast and withstand high voltages and high temperatures. SiC-based devices are ahead of the curve in terms of product availability in the market. GaN devices are gaining momentum in high-speed applications such as RF power amplifiers.

Unit 2 Overview of Power Semiconductor Devices

Text A Silicon Carbide Power Semiconductor Devices

Power semiconductor devices are attracting increasing attention as key components in a variety of power electronic systems. The major applications of power devices include power supplies, motor controls, renewable energy, transportation, telecommunications, heating, robotics, and electric utility transmission/distribution. The utilization of semiconductor power devices in these systems can enable significant energy savings, increased conservation of fossil fuels, and reduced environmental pollution.

Power electronics has gained renewed attention in the past decade due to the emergence of several new markets, including converters for photovoltaic and fuel cells, converters and inverters for electric vehicles (EVs) and hybrid-electric vehicles (HEVs), and controls for smart electric utility distribution grids. Currently, semiconductor power devices are one of the key enablers for global energy savings and electric power management in the future.

Silicon power devices have improved significantly over the past several decades, but these devices are now approaching performance limits imposed by the fundamental material properties of silicon, and further progress can only be made by migrating to more robust semiconductors. Silicon carbide (SiC) is a wide-bandgap semiconductor with superior physical and electrical properties that can serve as the basis for the high-voltage, low-loss power electronics of the future.

SiC is a IV-IV compound semiconductor with a bandgap of 2.3~3.3 eV (depending on the crystal structure, or polytype). It exhibits about 10 times higher breakdown electric field strength and 3 times higher thermal conductivity than silicon, making it especially attractive for high-power and high-temperature devices. For example, the on-state resistance of SiC power devices is orders-of-magnitude lower than that of silicon devices at a given blocking voltage, leading to much higher efficiency in electric power conversion. The wide bandgap and high thermal stability make it possible to operate certain types of SiC devices at junction temperatures of 300℃ or higher for indefinite periods without measurable degradation. Among wide-bandgap semiconductors, SiC is exceptional because it can be easily doped either p-type or n-type over a wide range, more than five orders-of-magnitude. In addition, SiC is the only compound semiconductor whose native oxide is SiO_2, the same insulator as silicon. This makes it possible to fabricate the entire family of MOS-based (metal-oxide-semiconductor) electronic devices in SiC.

Since the 1980s, sustained efforts have been directed toward developing SiC material and device technology. Based on a number of breakthroughs in the 1980s and 1990s, SiC

Schottky barrier diodes (SBDs) were released as commercial products in 2001. The market for SiC SBDs has grown rapidly over the last several years. SBDs are employed in a variety of power systems, including switch-mode power supplies, photovoltaic converters, air conditioners, and motor controls for elevators and subways. Commercial production of SiC power switching devices, primarily JFETs (junction field-effect transistors) and MOSFETs (metal-oxide-semiconductor field-effect transistors), began in 2006-2010. These devices are well accepted by the markets and many industries are now taking advantage of the benefits of SiC power switches. As an example, the volume and weight of a power supply or inverter can be reduced by a factor of 4～10, depending on the extent to which SiC components are employed. In addition to the size and weight reduction, there is also a substantial reduction in power dissipation, leading to improved efficiency in electric power conversion systems due to the use of SiC components.

Words and Phases

Power semiconductor device 功率半导体器件
power supply 电源
fuel cell 燃料电池
wide-bandgap *n.* 宽禁带
compound *adj.* 复合的
polytype *n.* 多型体
on-state voltage 导通电压
blocking voltage 阻断电压
junction temperature 结温
native oxide 本征氧化层
Schottky barrier diode 肖特基势垒二极管
photovoltaic converter 光伏变换器
metal-oxide-semiconductor field-effect transistor 金属氧化物半导体场效应晶体管

Note

1. Silicon power devices have improved significantly over the past several decades, but these devices are now approaching performance limits imposed by the fundamental material properties of silicon, and further progress can only be made by migrating to more robust semiconductors.

过去几十年，硅基功率器件的性能获得大幅提升，但受硅材料特性的制约，硅基功率器件性能已发挥到极限，为了进一步提升功率器件的性能，需要采用性能更加优异的半导体材料。

2. It exhibits about 10 times higher breakdown electric field strength and 3 times higher thermal conductivity than silicon, making it especially attractive for high-power and high-temperature devices.

与硅相比，它（碳化硅）具有高于10倍的击穿电场强度和高于三倍的热导率，这使得它（碳化硅）对制备高功率、高温器件更具吸引力。

3. In addition to the size and weight reduction, there is also a substantial reduction in power dissipation, leading to improved efficiency in electric power conversion systems due to

the use of SiC components.

除了体积和重量的减少，碳化硅器件的使用还可使功率损耗大幅降低，从而提升电能变换系统的效率。

Exercises

1. Answer the following questions according to the text

(1) What is the most common application of power semiconductor devices?

(2) What advanced properties does SiC possess compared with Si?

(3) Why can SiC devices operate at higher junction temperature?

(4) What is the mechanism behind the reduction of the size and weight of a power converter using SiC power devices?

2. Translate the following sentences intoChinese according to the text

(1) The major applications of power devices include power supplies, motor controls, renewable energy, transportation, telecommunications, heating, robotics, and electric utility transmission/distribution.

(2) The utilization of semiconductor power devices in these systems can enable significant energy savings, increased conservation of fossil fuels, and reduced environmental pollution.

(3) Silicon carbide (SiC) is a wide-bandgap semiconductor with superior physical and electrical properties that can serve as the basis for the high-voltage, low-loss power electronics of the future.

(4) As an example, the volume and weight of a power supply or inverter can be reduced by a factor of 4～10, depending on the extent to which SiC components are employed.

3. Translate the following intoChinese

The wide bandgap and high thermal stability make it possible to operate certain types of SiC devices at junction temperatures of 300℃ or higher for indefinite periods without measurable degradation. Among wide-bandgap semiconductors, SiC is exceptional because it can be easily doped either p-type or n-type over a wide range, more than five orders-of-magnitude. In addition, SiC is the only compound semiconductor whose native oxide is SiO_2, the same insulator as silicon. This makes it possible to fabricate the entire family of MOS-based (metal-oxide-semiconductor) electronic devices in SiC.

Since the 1980s, sustained efforts have been directed toward developing SiC material and device technology. Based on a number of breakthroughs in the 1980s and 1990s, SiC Schottky barrier diodes (SBDs) were released as commercial products in 2001. The market for SiC SBDs has grown rapidly over the last several years. SBDs are employed in a variety of power systems, including switch-mode power supplies, photovoltaic converters, air conditioners, and motor controls for elevators and subways. Commercial production of SiC power switching devices, primarily JFETs (junction field-effect transistors) and MOSFETs (metal-oxide-semiconductor field-effect transistors), began in 2006-2010. These devices are well accepted by the markets and many industries are now taking advantage of the benefits of SiC power switches.

Text B Introduction of SiC Devices in Comparison to Si Devices

At the heart of modern power electronics converters are power semiconductors switching devices. Today's power semiconductor devices are dominated by the mature and well established silicon (Si) technology. Since the advent of Si thyristors in 1957, many Si based switching devices have been developed to meet different application and performance needs. The most popular Si switching devices are insulated-gate bipolar transistors (IGBT) and power metal-oxide-field-effect transistors (MOSFET), with IGBT for high voltage, high power, and low frequency applications, and MOSFET for low voltage, low power and high frequency applications. Thyristors and their derivatives such as integrated-gate-commutated thyristors (IGCT) are still used in special high power applications.

Si power semiconductor devices have gone through many generations of development in the last 50 years and are approaching material theoretical limitations in terms of blocking voltage, operation temperature, and conduction and switching characteristics. Due to limited performance, the highest voltage rating of the state-of-art commercial Si IGBT has been 6.5 kV for the last 15years. There are no commercial Si based devices with junction temperature capability above 175℃. These intrinsic physical limits become a barrier to achieving higher performance power conversion.

The emergence of wide bandgap (WBG) semiconductor devices promises to revolutionize next-generation power electronics converters. Compared with Si devices, WBG devices feature high breakdown electric field, low specific on-resistance, fast switching speed and high junction temperature capability. All of these characteristics are beneficial for the efficiency, power density, specific power, and/or reliability of power electronics converters. The WBG devices under rapid development and commercialization include silicon carbide (SiC) and gallium nitride (GaN) devices, with SiC mainly targeting high voltage high power (600 V, kilowatts or above) applications, and GaN for low voltage low power (600 V, kilowatts or below) applications. SiC devices can improve and impact power electronics in several ways: ①At converter level, through substituting Si devices directly or simplifying circuit topologies, SiC devices can improve converter efficiency, reduce cooling needs, and reduce active and passive component numbers and size, with their high voltage, low loss and fast switching capabilities; ②At system level, SiC based converters can have better dynamic performance and more system functionalities as a result of their high frequency capability and high control bandwidth enabled by fast switching speed; ③SiC can enable new applications, such as high-efficiency high density solid-state transformers (SST) and high-speed motor drives. A number of commercial and research prototype converters using SiC devices have been developed with promising results on significantly improved efficiency and power density.

The extremely fast switching and other superior characteristics of SiC devices have nonetheless also posed severe challenges to their applications. Pervasive dv/dt and di/dt slew rates of up to 100 V/ns and 10 A/ns, augmented electromagnetic interference (EMI) emissions, single-device blocking voltages as high as tens of kV with corresponding insulation requirements, switching frequencies in the 100 s of kHz range, and junction temperatures surpassing 200℃, have called for a comprehensive reformulation of design procedures developed for Si-based power electronics. Addressing these design and application issues are critical to the adoption and success of SiC power electronics.

WBG refers to electronic energy band gaps significantly larger than one electron-volt (eV). SiC materials have several characteristics that make them attractive compared to narrow bandgap Si for power electronics converters. Generally speaking, for SiC material, the energy gap, breakdown electric field, thermal conductivity, melting point, and electron velocity are all significantly higher. These characteristics allow SiC semiconductor based power devices to operate at much higher voltage, switching frequency and temperature than Si.

For example, with the breakdown field higher than that of Si, a thinner drift layer with a higher doping concentration can be used for SiC power devices at the same blocking voltage. For unipolar device such as Schottky diodes and MOSFETs, the combination of thinner blocking layer and higher doping concentration yields a lower specific on-resistance compared with Si majority carrier devices.

The fast switching-speed capability of SiC devices can be expected due to higher breakdown field and electron velocity. First, with lower on-resistance at the same breakdown voltage, a reduced chip size is achieved in SiC unipolar devices such as MOSFET. Considering the tradeoff between thinner drift region and smaller chip size, the junction capacitance of SiC MOSFETs is still lower than that of the Si counterparts, therefore the switching speed becomes faster. Second, minority carriers are swept out of the depletion region at the saturated drift velocity during the turn-off transient. The electron saturated drift velocity of SiC is higher than that of Si, leading to an increased switching speed of SiC devices.

Additionally, the excellent thermal conductivity allows SiC dissipated heat to be readily extracted from the device. Hence, a larger power can be handled by the device at a given junction temperature. Also, higher thermal conductivity together with wide bandgap makes it possible for SiC devices to work at high temperature.

In summary, SiC based power devices offer low specificon-state resistance, fast switching speed, and high operating temperature and voltage capabilities.

Words and Phases

integrated-gate-commutated thyristors (IGCT) 集成门级换流晶闸管

intrinsic *adj*. 固有的，本身的

solid-state transformer 固态变压器

prototype *n.* 原型，样机
drift layer 漂移层
doping concentration 掺杂浓度
unipolar device 单极型器件
depletion region 耗尽区
drift velocity 漂移速度

Note

1. The emergence of wide bandgap (WBG) semiconductor devices promises to revolutionize next-generation power electronics converters. Compared with Si devices, WBG devices feature high breakdown electric field, low specific on-resistance, fast switching speed and high junction temperature capability.

宽禁带半导体器件的出现有望引发下一代电力电子变换器的技术变革。与硅器件相比，宽禁带器件具有高击穿电场，低导通电阻，高开关速度和高耐温。

2. SiC materials have several characteristics that make them attractive compared to narrow bandgap Si for power electronics converters.

与窄带隙的硅材料相比，碳化硅材料的几种特性使其在电力电子装置中应用时更具吸引力。

3. Considering the tradeoff between thinner drift region and smaller chip size, the junction capacitance of SiC MOSFETs is still lower than that of the Si counterparts, therefore the switching speed becomes faster.

综合考虑较薄的漂移层和较小的芯片体积两个特性，碳化硅 MOSFET 的结电容仍然比硅基器件小，因此开关速度更快。

Exercises

1. Answer the following questions according to the text

(1) What is the meaning of wide bandgap?

(2) Why does SiC MOSFET have fast switching-speed capability compared with Si counterparts?

(3) What are the challenges when SiC devices are used to build a power converter?

(4) Why the power density of the power converter can be highly improved using SiC devices?

2. Translate the following sentences intoChinese according to the text

(1) The most popular Si switching devices are insulated-gate bipolar transistors (IGBT) and power metal-oxide-field-effect transistors (MOSFET), with IGBT for high voltage, high power, and low frequency applications, and MOSFET for low voltage, low power and high frequency applications.

(2) A number of commercial and research prototype converters using SiC devices have been developed with promising results on significantly improved efficiency and power density.

(3) The extremely fast switching and other superior characteristics of SiC devices have

nonetheless also posed severe challenges to their applications.

(4) Si power semiconductor devices have gone through many generations of development in the last 50 years and are approaching material theoretical limitations in terms of blocking voltage, operation temperature, and conduction and switching characteristics.

3. Translate the following intoChinese

The fast switching-speed capability of SiC devices can be expected due to higher breakdown field and electron velocity. First, with lower on-resistance at the same breakdown voltage, a reduced chip size is achieved in SiC unipolar devices such as MOSFET. Considering the tradeoff between thinner drift region and smaller chip size, the junction capacitance of SiC MOSFETs is still lower than that of the Si counterparts, therefore the switching speed becomes faster. Second, minority carriers are swept out of the depletion region at the saturated drift velocity during the turn-off transient. The electron saturated drift velocity of SiC is higher than that of Si, leading to an increased switching speed of SiC devices.

Unit 3 AC-DC Converters

Text A Phase-Controlled Rectifiers

In most power electronic applications, the power input is the form of a 50 Hz or 60 Hz sine wave AC voltage provided by the electric utility, that is first converted to a DC voltage. Increasingly, the trend is to use the inexpensive rectifiers with diodes to convert the input AC into DC in an uncontrolled manner. The diode rectifiers provide a fixed output voltage only. To obtain controlled output voltages, phase-controlled thyristors are used instead of diodes. The output voltage of thyristor rectifiers is varied by controlling the delay or firing angle of thyristors. A phase-controlled thyristor is turned on by applying a short pulse to its gate and turned off due to natural or line commutation; and in case of a highly inductive load, it is turned off by firing another thyristor of the rectifier during negative half-cycle of input voltage.

Three phase-controlled rectifiers are simple and less expensive; and the efficiency of these rectifiers is, in general, above 95%. Because these rectifiers convert form AC to DC, these controlled rectifiers are also called AC-DC converters and are used extensively in industrial applications, especially in variable-speed drives, ranging from fractional horsepower to megawatt level.

The phase-controlled converters can be classified into two types, depending on the input supply: ① single-phase converters, and ② three-phase converters. Each type can be subdivided into semiconverter, full converter, and dual converter. A semiconverter is a one-quadrant converter and it has one polarity of output voltage and current. A full converter is a two-quadrant converter and the polarity of its output voltage can be either positive or negative. However, the output current of full converter has one polarity only. A dual converter can operate in four quadrants; and both the output voltage and current can be either positive or negative. In some applications, converters are connected in series to operate at higher voltages and to improve the input power factor (PF).

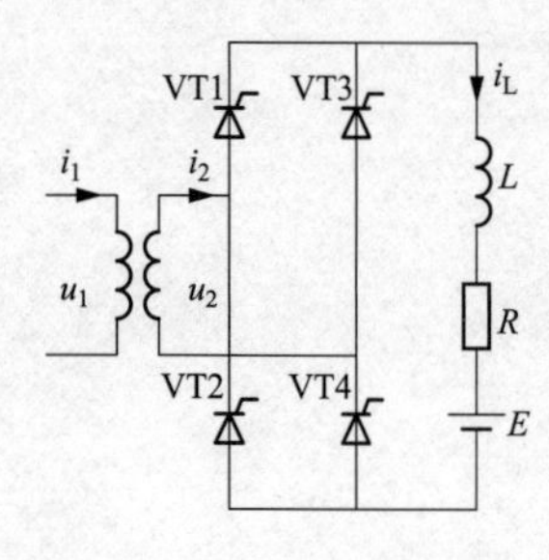

Fig. 3. 2 Single-phase full converter

Fig. 3. 2 shows the single-phase full converter with the resistance, inductance, and counter electromotive force (CEMF) load. The load inductance is assumed to be very large so that the load current i_L is ripple-free. In the positive half-cycle, thyristors VT1 and VT4 are forward-biased, and when these two devices are triggered into conduction at firing angle a, the load current will flow through these devices. Since the load is inductive, the current will continue to flow beyond π angle when the voltage reverses its polarity. Thyristors VT2 and VT3 are fired symmetrically at a angle in the next half-cycle. This places a reverse voltage across VT1 and VT4, which turn off, or commutates, and the load current is taken by VT2 and VT3. This method of commutation is known as natural. Each thyristor pair conducts for a half-cycle. The thyristor can be fired by single gate current pulse, but generally, pulse-train firing during the entire half-cycle is preferred so that the device does not turn off inadvertently if the anode current is fluctuating and falls below the holding current.

Words and Phases

electric utility 电力系统
firing angle 触发角
phase-controlled rectifier 相控整流器
variable-speed drive 变速驱动
fractional horsepower 小马力（低功率）
megawatt *n.* 兆瓦
semiconverter *n.* 半控变换器
full converter 全控变换器
dual converter 双变换器
quadrant *n.* 象限
counter electromotive force 反电动势
ripple-free *n.* 无波动的，无脉动的
forward-biased *adj.* 正向偏置的
symmetrically *adv.* 对称地
pulse-train *n.* 脉冲列

Notes

A phase-controlled thyristor is turned on by applying a short pulse to its gate and turned off due to natural or line commutation; and in case of a highly inductive load, it is turned off by firing another thyristor of the rectifier during negative half-cycle of input voltage.

相控晶闸管通过向其栅极施加短脉冲来开通，并通过自然或线路换向来关断；如果是高感性负载，则在输入电压的负半周期内通过触发整流器的另一个晶闸管将其关断。

Exercises

1. Answer the following questions according to the text

(1) What are the advantages of phase-controlled rectifiers?

(2) Whatis the semiconverter?

(3) Which commutation method does the single-phase full converter with a large inductive load use?

(4) Why the gate pulse-train firing is used during the entire half-cycle for the single-phase full converter?

2. Translate the following sentences into Chinese according to the text

(1) In most power electronic applications, the power input is the form of a 50 Hz or 60 Hz sine wave ac voltage provided by the electric utility, that is first converted to a DC voltage.

(2) The output voltage of thyristor rectifiers is varied by controlling the delay or firing angle of thyristors.

(3) In the positive half-cycle, thyristors VT1 and VT4 are forward-biased, and when these two devices are triggered into conduction at firing angle a, the load current will flow through these devices.

3. Translate the following into Chinese

Three phase-controlled rectifiers are simple and less expensive; and the efficiency of these rectifiers is, in general, above 95%. Because these rectifiers convert form AC to DC, these controlled rectifiers are also called AC-DC converters and are used extensively in industrial applications, especially in variable-speed drives, ranging from fractional horsepower to megawatt level.

Text B PWM Rectifier

There are several applications where energy flow can be reversed during the operation. Examples are: locomotives, downhill conveyors, cranes, etc. In all these applications, the line-side converter must be able to deliver energy back to the power supply, what is known as power regeneration. A pulse width modulation (PWM) rectifier with bidirectional power flow capability, sinusoidal grid current, and power factor control is the most common rectifier in these applications.

Fig. 3. 3 (a) shows the power circuit of the fully controlled single-phase PWM rectifier in bridge connection, which uses four controlled power switches with antiparallel diodes to produce a controlled DC voltage U_o. For appropriate operation of this rectifier, the output voltage must be greater than the input voltage, at any time ($U_o > \tilde{u}_s$). This rectifier can work with two (bipolar PWM) or three (unipolar PWM) levels as shown in Fig. 3. 3.

The possible combinations are as follows.

➢ Switch VT1 and VT4 are in ON state and VT2 and VT3 are in OFF state, $u_{AFE}=U_o$ [Fig. 3. 3 (b)].

➢ Switch VT1 and VT4 are in OFF state and VT2 and VT3 are in ON state, $u_{AFE}=U_o$ [Fig. 3. 3 (c)].

➢ Switch VT1 and VT3 are in ON state and VT2 and VT4 are in OFF state, or VT1 and VT3 are in OFF state and VT2 and VT4 are in ON state, $u_{AFE}=0$ [Fig. 3. 3 (d)].

The inductor voltage can be expressed as

$$u_L = L\frac{di_s}{dt} = v_s(t)-kU_o$$

where $k=1$, -1 or 0.

If $k=1$, then the inductor voltage will be negative, so the input current i_s will decrease its value.

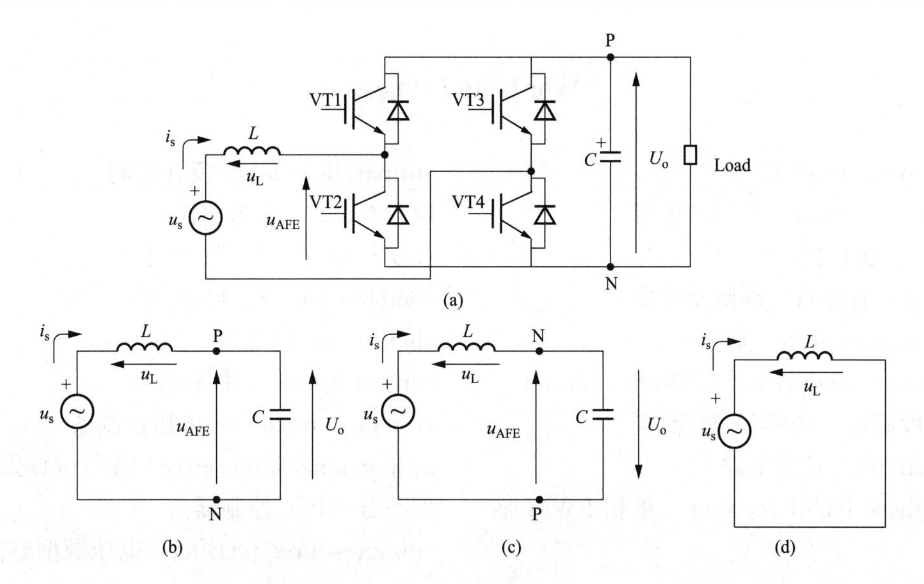

Fig. 3. 3 Single-phase PWM rectifier in bridge connection

(a) Power circuit; (b) Equivalent circuit with VT1 and VT4 ON;

(c) VT2 and VT3 ON; (d) VT1 and VT3 or VT2 and VT4 ON

If $k=-1$, then the inductor voltage will be positive, so the input current i_s will increase its value.

Finally, if $k=0$ the input current increase or decrease its value depending of u_s. This allows for a complete control of the input current.

If condition $U_o>\tilde{u}_s$ is not satisfied, for example during startup, the input current cannot be controlled and the capacitor will be charged through the diodes to the peak value of the source voltage ($\tilde{u}_s$) as a typical noncontrolled rectifier. After that, the converter will start working in controlled mode increasing the output voltage U_o to the reference value.

The classical control scheme is shown in Fig. 3. 4. The control includes a voltage controller, typically a proportional-integrative (PI) controller, which controls the amount of power required to maintain the DC-link voltage constant. The voltage controller delivers the amplitude of the input current. For this reason, the voltage controller output is multiplied by a sinusoidal signal with the same phase and frequency than u_s, in order to obtain the input current reference, i_{sref}. The fast current controller controls the input current, so the high input power factor is achieved. Note that for PWM operation the voltage-source rectifier (VSR) must have a capacitive filter at the DC side and an inductive filter at the AC side.

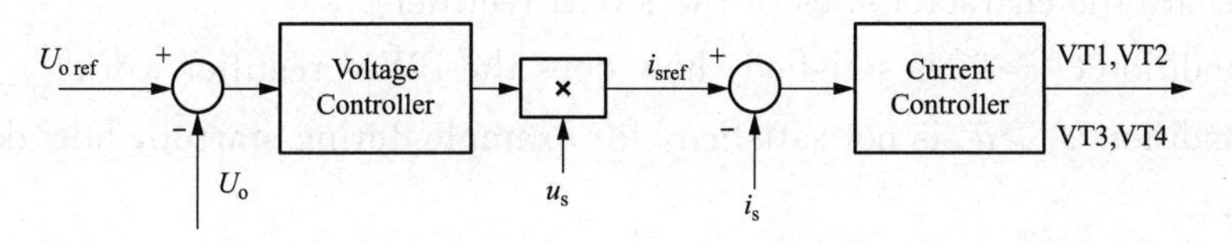

Fig. 3. 4 Control scheme of bridge PWM rectifier

Words and Phases

locomotive *n.* 机车
downhill conveyor 下运输送机
crane *n.* 起重机
line-side converter 网侧变流器
power regeneration 能量再生
pulse width modulation (PWM) rectifier 脉冲宽度调制（PWM）整流器
power circuit 功率电路
single-phase PWM rectifier 单相 PWM 整流器
antiparallel *adj.* 反并联的
DC voltage 直流电压
two/three level 两/三电平
combination *n.* 组合
charge *v.* 使充电
control scheme 控制方案
voltage controller 电压控制器
proportional-integrative (PI) controller 比例积分（PI）控制器
voltage-source rectifier 电压源型整流器

Notes

1. For appropriate operation of this rectifier, the output voltage must be greater than the input voltage, at any time ($U_o > \tilde{u}_s$). This rectifier can work with two (bipolar PWM) or three (unipolar PWM) levels as shown in Fig. 3. 3.

为了适当操作该整流器，输出电压必须大于输入电压（$U_o > \tilde{u}_s$）。该整流器可以工作在两（双极性 PWM）或三（单极性 PWM）电平，如图 3. 3 所示。

2. If condition $U_o > \tilde{u}_s$ is not satisfied, for example during startup, the input current cannot be controlled and the capacitor will be charged through the diodes to the peak value of the source voltage ($\tilde{u}_s$) as a typical noncontrolled rectifier. After that, the converter will start working in controlled mode increasing the output voltage u_o to the reference value.

如果条件 $U_o > \tilde{u}_s$ 不满足，例如在启动期间，输入电流无法控制，电容器将通过二极管充电到输入交流的峰值电压（$\tilde{u}_s$），此时 PWM 整流器作为典型的不控整流器运行。之后，整流器将开始以可控模式工作，将输出电压 u_o 增加到参考值。

Exercises

1. Answer the following questions according to the text

(1) What are the characteristics of the PWM rectifier?

(2) If condition $U_o > \tilde{u}_s$ is satisfied, how does the PWM rectifier work?

(3) If condition $U_o > \tilde{u}_s$ is not satisfied, for example during startup, how does the PWM rectifier work?

(4) Please describe the control scheme of the PWM rectifier.

2. Translate the following sentences into Chinese according to the text

(1) In all these applications, the line-side converter must be able to deliver energy back

to the power supply, what is known as power regeneration.

(2) Fig. 3. 3 (a) shows the power circuit of the fully controlled single-phase PWM rectifier in bridge connection, which uses four controlled power switches with antiparallel diodes to produce a controlled dc voltage U_o.

(3) Finally, if $k=0$ the input current increase or decrease its value depending of u_s. This allows for a complete control of the input current.

3. Translate the following paragraph into Chinese

The classical control scheme is shown in Fig. 3. 4. The control includes a voltage controller, typically a proportional-integrative (PI) controller, which controls the amount of power required to maintain the DC-link voltage constant. The voltage controller delivers the amplitude of the input current. For this reason, the voltage controller output is multiplied by a sinusoidal signal with the same phase and frequency than u_s, in order to obtain the input current reference, i_{sref}. The fast current controller controls the input current, so the high input power factor is achieved. Note that for PWM operation the voltage-source rectifier (VSR) must have a capacitive filter at the DC side and an inductive filter at the AC side.

Unit 4 DC-DC Converter

Text A Trend for Distributed Power Systems

Power electronics products, to date, are essentially custom-designed. With a long design cycle, today's power electronics equipment is designed and manufactured using non-standard parts. Thus, manufacturing processes are labor-intensive, resulting in high cost and poor reliability. This practice has significantly weakened the U. S. power electronics industry in recent years. In the 1980s, power electronics was considered to be a core enabling technology for all of the major corporations in the U. S. In the 1990s, the major corporations adopted an outsourcing strategy and spun off their power electronics divisions. What had been a captive market was transformed into a merchant market. Fewer resources were available to devote to technical advancements in power electronics. Therefore, innovative solutions were scarce, and products became commoditized and cost-driven. The result was the increased mass migration of manufacturing to countries with low labor costs. The problem is further compounded by the more recent trend of outsourcing engineering to India and Asian countries, especially China. Today, most of the industry is focused on the bottom line and spends little on R and D, and of this most is spent on development rather than on research.

With ever-increasing current consumption (>100 A) and clock frequency (>GHz), today's microprocessors are operating at very low voltages (1 V or less) and continuously switching between the sleep-mode and wake-up mode at frequencies of up to several MHz to conserve energy. Lee has proposed a multiphase voltage regulator (VR) module for new generations of Intel Pentium microprocessors (Fig. 3. 5).

This multiphase VR technology is simple and is easily scalable to meet ever-increasing current consumption, clock rates, and stringent voltage regulation requirements. With the ability to adjust phases according to the load demand, it demonstrates superior efficiency at light load, at which microprocessors operate most of the time. Today, every PC and server microprocessor in the world is powered with this VR. These technologies have been further extended to high-performance graphical processors, server chipset and memory devices, networks, telecommunications, and all forms of mobile electronics. For example, in telecommunication applications, 48 V has been adopted as the low-voltage bus for all switch boards. A number of isolated DC/DC converters, often referred to as brick converters, are employed for conversion from 48 V to low-voltage outputs in each switch board. The brick converters demand high power density and high efficiency, and thus are very competitive products and laden with IPs. With the advent of multiphase VRs, these brick products have been quickly converted into a two-stage solution, with a simple DC/DC transformer (DCX) followed by a multiphase VR, as shown in Fig. 3. 6, with demonstrated improvement in efficiency, power density, and even cost. With the widespread use of distributed point-of-Load, the power architecture has undergone significant changes, from centralized customized power supplies to more distributed power supplies (DPS), as shown in Fig. 3. 7. In this DPS architecture, the modular approach has been adopted for even the front-end power conversion. The DPS architecture offers many benefits, such as modularity, scalability, fault tolerance, improved reliability, serviceability, redundancy, and a reduced time to market.

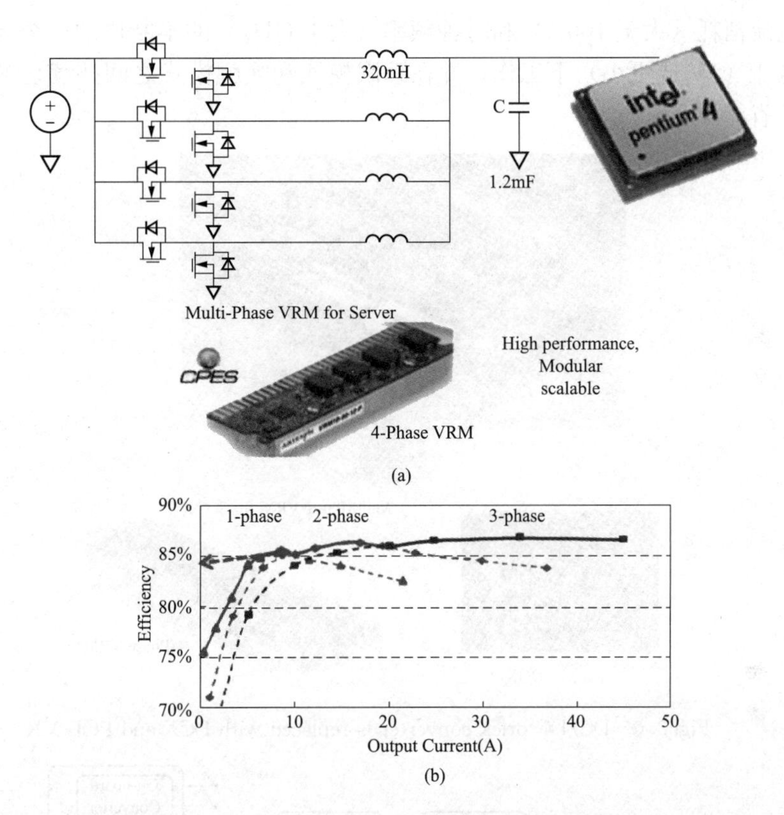

Fig. 3. 5 A multiphase voltage regulatot (VR) module for new generation of Intel Pentium microprocessors
(a) Multiphase VR for the new generation of microprocessors; (b) Efficiency improvement at light load using phase shedding and burst-mode operation

Words and Phases

multiphase *n.* 多相
centralized *adj.* 集中式的
custom-designed *adj.* 定制的
outsourcing *adj.* 外包的
chipset *n.* 芯片集
switch boards 开关板
architecture *n.* 结构
modularity *n.* 模块性
fault tolerance 容错性
redundancy *n.* 冗余性

Notes

1. With ever-increasing current consumption (>100 A) and clock frequency (>GHz), today's microprocessors are operating at very low voltages (1 V or less) and continuously switching between the sleep-mode and wake-up mode at frequencies of up to several MHz to conserve energy.

随着电流消耗（大于100A）和时钟频率（大于GHz）的不断增加，今天的微处理器在非常低的电压（1V或更小）下工作，并在睡眠模式和唤醒模式之间持续切换，频率高达几兆赫，以节省能源。

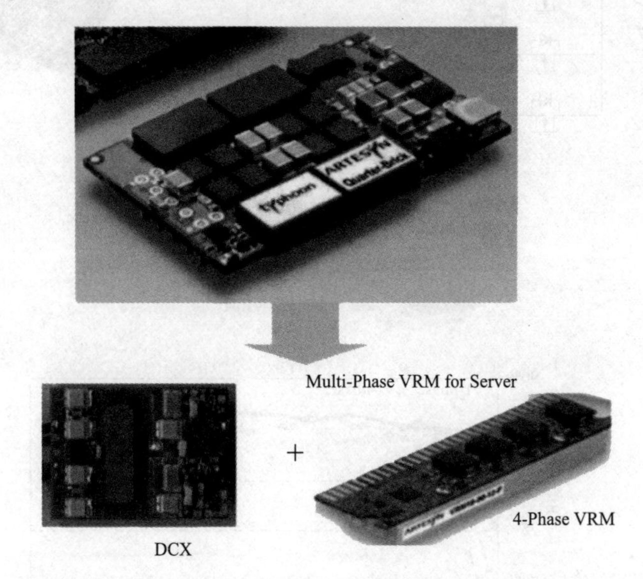

Fig. 3. 6 DC/DC brick converter is replaced with DCX and POL VR

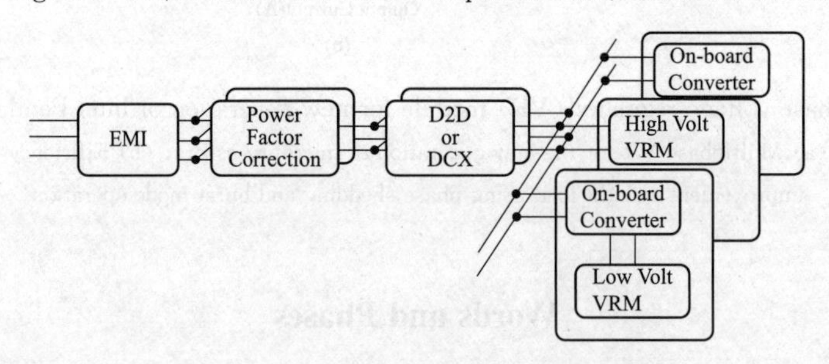

Fig. 3. 7 Distributed power architecture

2. With the widespread use of distributed point-of-Load, the power architecture has undergone significant changes, from centralized customized power supplies to more distributed power supplies (DPS).

随着分布式负载点的广泛使用，电源结构发生了重大变化，从集中式定制电源到更加分布式的电源（DPS）。

Exercises

1. Answer the following questions according to the text

(1) Why is the U. S. power and electronics industry weakened?

(2) What are the reasons why products in most industries are becoming commercialized and cost-driven today?

(3) What are the characteristics of multi-phase virtual reality technology?

(4) What major changes have taken place in the power supply structure?

(5) What are the advantages of DPS architecture?

2. Translate the following sentences into Chinese according to the text

(1) In the 1990s, the major corporations adopted an outsourcing strategy and spun off their power electronics divisions.

(2) Today, most of the industry is focused on the bottom line and spends little on R and D, and of this most is spent on development rather than on research.

(3) This multiphase VR technology is simple and is easily scalable to meet ever-increasing current consumption, clock rates, and stringent voltage regulation requirements.

(4) These technologies have been further extended to high-performance graphical processors, server chipset and memory devices, networks, telecommunications, and all forms of mobile electronics.

(5) The DPS architecture offers many benefits, such as modularity, scalability, fault tolerance, improved reliability, serviceability, redundancy, and a reduced time to market.

3. Translate the following into Chinese

With the continuous progress of semiconductor technology and the development of integrated technology, the characteristic size of microprocessor chips is getting smaller and smaller. Because of the process requirements, and also in order to obtain the lowest power consumption and the highest efficiency, the power supply voltage of bit processors tends to be low-voltage. With the decrease of the working voltage of bit processors, their working voltage becomes smaller and smaller. The requirement of stability becomes higher. Voltage fluctuation of 50 mV may lead to wrong operation of the circuit. At the same time, the stronger the function of microprocessor, the larger the circuit size and internal function circuit, and the larger the driving current provided by the power supply. At the same time, microprocessor needs to work according to different work in order to reduce power consumption. In order to maintain high voltage stability under the condition of low voltage, high current and high current change rate, the power supply module of microprocessor is required to have excellent dynamic performance.

Text B DC-DC Converter in Photovoltaic System

High step-up Boost converters with coupled inductor have attracted much attention in the fuel cell or photovoltaic (PV) grid-connected generation system, however, there are few literatures elaborated on the construction ideas and derivation methods of them. Accordingly, in order to obtain a clear roadmap on the derivation and inner connection of these converters, a comprehensive review and analysis are presented in this paper. Firstly, basic Boost converter with coupled inductor is regarded as the basic topology, and its merits and demerits are analyzed in detail. Then, in order to address these demerits, various step-up techniques are introduced, such as the rectifier circuit, the active-clamped circuit, the multi-winding coupled inductor and the voltage doubler rectifier; and numerous new topologies are continuously proposed by combinations and equivalent simplifications. In addition to a detailed synthesis of each topology, a comparative and quantitative analysis among some important converters are presented, and the optimal one is chosen to build a 250 W prototype. Finally, based on comparisons and analysis, the main characteristics and inner connections of these high step-up Boost converters with coupled inductor are identified and clarified.

At present, fossil fuels (such as coal, petroleum, and natural gas) are still the main sources of energy consumption, and thermal power is the most important way to generate electricity in the world. However, the energy crisis and environmental problems have become increasingly serious. In order to reduce the proportion of fossil energy in the energy consume structure and resolve environmental problems, the development of renewable energy is inevitable. For current renewable energy resources, solar energy and biomass energy are of great significance due to their characteristics of wide distribution, no geographical restrictions and pollution-free. In a grid-connected generation system, which includes photovoltaic panels and fuel cells, the output voltages of the photovoltaic arrays and fuel cells are commonly range from 20 V to 50 V; apparently, the output voltage is far less than the required voltage of a grid-connected inverter, therefore, high step-up DC-DC converter is necessary in a grid connected inverter system.

The conventional Boost converter is a standard non-isolated step-up converter, and theoretically, when the duty cycle approaches to one, the voltage gain tends to infinity. However, large duty cycle causes the problems of large voltage stresses on switches, serious diode reverse recovery, low efficiency and so on. Furthermore, due to the influence of parasitic parameters, the voltage gain even decreases with the increase of duty cycle, and it cannot satisfy the requirements of the abovementioned high step-up DC-DC conversion applications.

To overcome the limitations that imposed by extreme duty cycle, many kinds of high step-up DC-DC converters have been proposed; and these converters can be divided into three categories: ①the cascaded DC-DC converter, where the output of the former stage converter is also the input of the lager stage converter, and the system voltage gain equals to the product of these two converters. However, there are some shortcomings in this kind of converter, such as complicate control system, numerous components and high voltage stresses on the later stage converter; ② the switched capacitor DC-DC converter , to achieve high step-up ratio, a combination of multiple switched capacitor units is required, and the voltage stress on each cell is reduced. However, only an integral multiple of voltage gain can be achieved. Moreover, since the capacitors are operating in switching state, there will be a large spike current; therefore, it is generally used for small power applications; ③the coupled inductor DC-DC converter, by introducing a coupled inductor, the system voltage gain can be improved by reasonable design the turns ratio of the coupled inductor, and it is easy to achieve high step-up conversion.

The high step-up DC-DC converters for photovoltaic grid connected system are classified in , and the performance of each converter is analyzed, but the inner connection among different converters are not revealed. Coupled inductor converter has the characteristics of simple structure, low cost and easy to realize high step-up conversion, therefore, it has been widely adopted, and various kinds of coupled inductor Boost converters are proposed . There are significant differences among the circuit topologies, however, their performances are similar, and the inner connection among them is not clear. Therefore, it is necessary to induce, summary and clarify the methods and ideas to derive these topologies. A review is made on the existing coupled inductor high step-up Boost converters, and the merits and faults of each converter are given. Moreover, it tries to explain the inner connection among them, but it is not systematic, complete, clear and accurate. A comprehensive description and comparative analysis of high step-up DC-DC converters adopting coupled inductor and voltage multiplier cells are given in, the main characteristics and constraints of the analyzed converters are identified, and however, the inner connection between these topologies is still not mentioned.

In this paper, starting with the basic coupled inductor Boost converter, its merits and demerits are analyzed; then, in order to overcome its demerits, different kinds of high step-up Boost converters with coupled inductor are derived based on various techniques. By classifying, deriving and analyzing these converters, the characteristics of each converter are clearly expounded and the inner connections among them are clarified.

Words and Phases

photovoltaic *adj*. 光伏
elaborate *v*. 阐述
converter *n*. 转换器
renewable energy 再生能源
coupled inductor 耦合电感
topology *n*. 拓扑
reveal *v*. 揭示
accurate *adj*. 精确的
fossil fuel 化石燃料
thermal power 火力发电
photovoltaic array 光伏阵列
non-isolated step-up converter 非隔离升压转换器
duty cycle 占空比
parasitic parameter 寄生参数
systematic *adj*. 系统的
the cascaded DC-DC converter 级联 DC-DC 转换器
the switched capacitor DC-DC converter 开关电容器 DC-DC 转换器
the coupled inductor DC-DC converter 耦合电感 DC-DC 转换器
high step-up Boost converters 高升压变压器

Notes

1. In addition to a detailed synthesis of each topology, a comparative and quantitative analysis among some important converters are presented, and the optimal one is chosen to build a 250 W prototype.

除了详细分析每种拓扑外，还进行了一些重要转换器的比较和定量分析，并选择最佳的转换器来构建 250 W 样机。

2. The conventional Boost converter is a standard non-isolated step-up converter, and theoretically, when the duty cycle approaches to one, the voltage gain tends to infinity. However, large duty cycle causes the problems of large voltage stresses on switches, serious diode reverse recovery, low efficiency and so on.

传统的 Boost 转换器是标准的非隔离升压转换器，理论上，当占空比接近 1 时，电压增益趋于无穷大。然而，大占空比导致开关上的电压应力大，二极管反向恢复严重，效率低等问题。

Exercises

1. Answer the following questions according to the text

(1) Why we need a high step-up DC-DC converter in a grid connected inverter system?

(2) What will happen when a conventional Boost converter duty cycle approaches to one?

(3) How many kinds of high step-up DC-DC converters? What are they?

(4) What are the characteristics of the coupled inductor converter?

(5) What is the role of introducing a coupled inductor DC-DC converter?

2. Translate the following sentences into Chinese according to the text

(1) High step-up Boost converters with coupled inductor have attracted much attention in the fuel cell or photovoltaic (PV) grid-connected generation system.

(2) At present, fossil fuels (such as coal, petroleum, and natural gas) are still the main sources of energy consumption, and thermal power is the most important way to generate electricity in the world.

(3) In a grid-connected generation system, which includes photovoltaic panels and fuel cells, the output voltages of the photovoltaic arrays and fuel cells are commonly range from 20 V to 50 V.

(4) Coupled inductor converter has the characteristics of simple structure, low cost and easy to realize high step-up conversion, therefore, it has been widely adopted, and various kinds of coupled inductor Boost converters are proposed.

(5) The high step-up DC-DC converters for photovoltaic grid connected system are classified in, and the performance of each converter is analyzed.

3. Translate the following into Chinese

To overcome the limitations that imposed by extreme duty cycle, many kinds of high step-up DC-DC converters have been proposed; and these converters can be divided into three categories: ①the cascaded DC-DC converter , where the output of the former stage converter is also the input of the lager stage converter, and the system voltage gain equals to the product of these two converters. However, there are some shortcomings in this kind of converter, such as complicate control system, numerous components and high voltage stresses on the later stage converter; ②the switched capacitor DC-DC converter , to achieve high step-up ratio, a combination of multiple switched capacitor units is required, and the voltage stress on each cell is reduced. However, only an integral multiple of voltage gain can be achieved. Moreover, since the capacitors are operating in switching state, there will be a large spike current; therefore, it is generally used for small power applications; ③ the coupled inductor DC-DC converter, by introducing a coupled inductor, the system voltage gain can be improved by reasonable design the turns ratio of the coupled inductor, and it is easy to achieve high step-up conversion.

Unit 5 DC-AC Converter

Text A Transformerless Inverters

In photovoltaic (PV) applications, a transformer is often used to provide galvanic isolation and voltage ratio transformations between input and output. However, these conventional iron and copper-based transformers increase the weight/size and cost of the inverter whilst reducing the efficiency and power density. It is therefore desirable to avoid using transformers in the inverter. However, additional care must be taken to avoid safety hazards such as ground fault currents and leakage currents, e. g. via the parasitic capacitor between the PV panel and ground. Consequently, the grid connected transformerless PV inverters must comply with strict safety standards such as interconnection and interoperability requirements standard for relevant interfaces between interconnected distributed energy and power system IEEE 1547.1, standard requirements for automatic disconnection device between generator and public low voltage grid VDE0126-1-1, safety standard for household and similar electrical appliances EN 50106, standard for characteristics of the utility interface in PV systems IEC61727, and the latest standards for the installation of solar panels AS/NZS 5033.

Various transformerless inverters have been proposed recently to eliminate the leakage current using different techniques such as decoupling the DC from the AC side and/or clamping the common mode (CM) voltage during the freewheeling period, or using common ground configurations. The permutations and combinations of various decoupling techniques with integrated voltage buck-boost for maximum uepower point tracking (MPPT) allow numerous new topologies and configurations which are often confusing and difficult to follow when seeking to select the right topology. Therefore, to present a clear picture on the development of transformerless inverters for the next generation grid-connected PV systems, this paper aims to comprehensively review and classify various transformerless inverters with detailed analytical comparisons. To reinforce the findings and comparisons as well as to give more insight on the CM characteristics and leakage current, computer simulations of major transformerless inverter topologies have been performed in PLECS software. Moreover, the cost and size are analyzed properly and summarized in a table. Finally, efficiency and thermal analysis are provided with a general summary as well as a technology roadmap.

Solar photovoltaic (PV) is one of the cleanest, readily and widely available energy sources among all renewable energies. With technological advancements in material and manufacturing techniques, the cost of the PV system is continuously reduced, making it the cheapest energy source for massive deployment in the future. Many countries (USA, Germany, China, Japan, Australia, France, Italy, Spain, etc.) have already begun to reap the benefits through their increased adoption and integration of this system in the utility grid. According to the 2017 annual report of the International Energy Agency-Photovoltaic Power Systems Program (IEA-PVPS), the global installed PV capacity reached a 100 GW milestone in 2012, and a 200 GW level in 2015. By the end of 2017, the total installed PV capacity was estimated to be roughly 410 GW, while 24 IEA-PVPS countries reached 264 GW. Fig. 3. 8 shows the cumulative installed PV capacity of the top IEA-PVPS countries from 2012 to 2017. From this figure, it is evident that the PV industry is facing rapid growth with five leading countries representing 90.1% of all PV installations in 2017. Among them, China, USA, and Japan experienced the largest installed PV installation capacity increment in recent years.

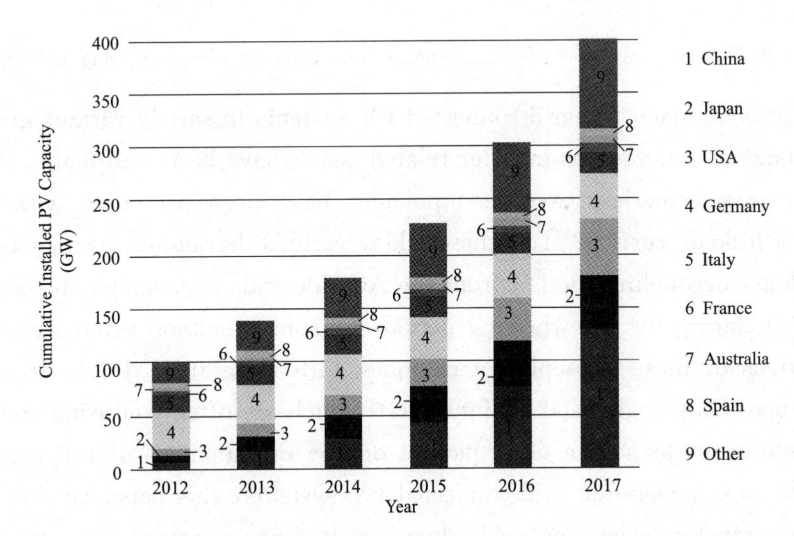

Fig. 3. 8 Cumulative PV installations for the top IEA-PVPS countries from 2012 to 2017

Among all PV installations, the percentage of off-grid PV systems is very low. The grid-connected PV systems need power inverters as interfaces between the PV panel and the grid, and these are generally categorized as galvanic isolated inverters and non-isolated inverters. In the isolated type, usually a high-frequency DC side transformer or a low-frequency AC side transformer is used to achieve galvanic isolation and this serves to enhance the overall system safety. Due to their lower cost, size/weight, and higher efficiency, transformerless inverters have generated a high degree of interest in terms of residential market with low to medium power capacity.

Fig. 3. 9 illustrates a general layout for a single-phase transformerless inverter for small-scale PV systems. As can be seen, without a galvanic isolation, a direct ground-current path may form between the PV panel and the grid. Due to the presence of large stray capacitance (C_{PV}) between the PV and grid grounds, the varying voltage (also known as common-mode (CM) voltage) can excite the resonant circuit formed by the parasitic capacitor and inverter filter inductor and this produces a high CM ground current i_{dn}. This capacitive i_{dn} comprises line low-frequency and switching high-frequency components which inject harmonics into the grid current, increase the system losses, impair the electromagnetic compatibility, and can cause safety problems such as electric shock.

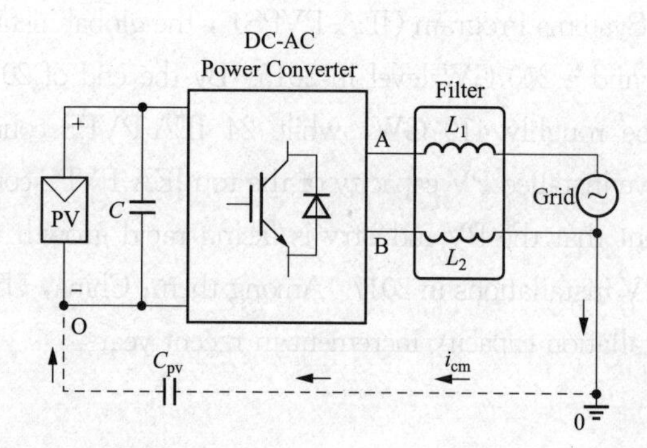

Fig. 3. 9 The general layout of a single-phase transformerless inverter using an L-filter

In order to understand the grid-connected PV systems to satisfy various grid codes and their safety standards, numerous inverter related issues have been thoroughly investigated. So far, many transformerless inverter topologies have been presented with the aim of eliminating the leakage current. To achieve this, various decoupling techniques have been adopted, such as, decoupling the DC from the AC side and/or clamping the common mode voltage (CMV) during the freewheeling period, or using common ground configurations. The combinations of these decoupling techniques with integrated MPPT circuits form an immense number of topologies and configurations which are often confusing and difficult to follow. Therefore, to present a clear picture on the development of the transformerless inverter for the next generation grid-connected PV systems, this paper aims to review and classify various transformerless inverters. Further, it aims to provide an analytical overview and analysis of well-known single-phase transformerless inverter topologies as well as a comparison of the transformerless inverters based on a loss and efficiency analysis through the means of detailed calculations. This categorisation and analysis can help researchers to understand the advantages and disadvantages of various transformerless inverter topologies in terms of their CMV and leakage current behaviour.

Words and Phases

galvanic isolation 电气隔离
ground fault currents 接地故障电流
leakage currents 漏电流
parasitic capacitor 寄生电容
freewheeling period 续流周期
cumulative *adj*. 累积的，渐增的
stray capacitance 杂散电容
resonant circuit 谐振电路
transformerless inverters 非隔离逆变器
maximum power point tracking (MPPT) 最大功率点跟踪

Notes

1. It is therefore desirable to avoid using transformers in the inverter. However, additional care must be taken to avoid safety hazards such as ground fault currents and leakage currents, e. g. via the parasitic capacitor between the PV panel and ground.

因此，最好避免在逆变器中使用变压器。但是，必须采取额外的注意措施，以避免诸如接地故障电流和漏电流等危险，例如光伏电池板和接地之间的寄生电容引入。

2. Various transformerless inverters have been proposed recently to eliminate the leakage current using different techniques such as decoupling the DC from the AC side and/or clamping the common mode (CM) voltage during the freewheeling period, or using common ground configurations.

为了消除漏电流，最近提出了各种无变压器逆变器，如将直流与交流侧在续流阶段解耦或钳位共模电压，或采用共同接地方式。

Exercises

1. Answer the following questions according to the text

(1) Why is it best to avoid using transformers in inverters ?

(2) There are several forms of transformerless inverter?

(3) Why are more and more countries paying attention to solar photovoltaic?

(4) What happens without current isolation in the overall layout of single-phase transformer-free inverters for small PV systems?

(5) What is the effect of voltage change caused by stray capacitance?

2. Translate the following sentences into Chinese according to the text

(1) The permutations and combinations of various decoupling techniques with integrated voltage buck-boost for maximum power point tracking (MPPT) allow numerous new topologies and configurations which are often confusing and difficult to follow when seeking to select the right topology.

(2) With technological advancements in material and manufacturing techniques, the cost of the PV system is continuously reduced, making it the cheapest energy source for massive deployment in the future.

(3) The grid-connected PV systems need power inverters as interfaces between the PV panel and the grid, and these are generally categorized as galvanic isolated inverters and non-isolated inverters.

(4) In the isolated type, usually a high-frequency DC side transformer or a low-frequency AC side transformer is used to achieve galvanic isolation and this serves to enhance the overall system safety.

(5) Due to the presence of large stray capacitance (C_{PV}) between the PV and grid

grounds, the varying voltage (also known as common-mode (CM) voltage) can excite the resonant circuit formed by the parasitic capacitor and inverter filter inductor and this produces a high CM ground current i_{dn}.

3. Translate the following into Chinese

Depending on PV modules configuration, inverters are mainly categorized as central, String, Multi-string and Module inverters. The central inverters are used for three phase large power applications while others String, Multi-string and Module inverters can be three or single phase depending on power rating and type of applications.

Furthermore, single phase PV inverters can be classified on the basis of galvanic isolation such as with-transformer andtransformerless. The detailed classification of grid interactive PV inverter topologies are illustrated in Fig. 3. 8.

Text B Evolution of PV Inverters

1. The Past—Centralized Inverters

The past technology was based oncentralized inverters that interfaced a large number of PV modules to the grid. The PV modules were divided into series connections (called a string), each generating a sufficiently high voltage to avoid further amplification. These series connections were then connected in parallel, through string diodes, in order to reach high power levels. This centralized inverter includes some severe limitations, such as high-voltage DC cables between the PV modules and the inverter, power losses due to a centralized MPPT, mismatch losses between the PV modules, losses in the string diodes, and a nonflexible design where the benefits of mass production could not be reached. The grid-connected stage was usually line commutated by means of thyristors, involving many current harmonics and poor power quality.

The large amount of harmonics was the occasion of new inverter topologies and system layouts, in order to cope with the emerging standards which also covered power quality.

2. The Present—String Inverters and AC Modules

The present technology consists of the string inverters and the AC module. The string inverter, is a reduced version of the centralized inverter, where a single string of PV modules is connected to the inverter. The input voltage may be high enough to avoid voltage amplification. This requires roughly 16 PV modules in series for European systems.

The total open-circuit voltage for 16 PV modules may reach as much as 720 V, which calls for a 1000 V MOSFET/IGBT in order to allow for a 75% voltage de-rating of the semiconductors. The normal operation voltage is, however, as low as 450～510 V. The possibility of using fewer PV modules in series also exists, if a DC-DC converter or line-frequency transformer is used for voltage amplification. There are no losses associated with string diodes and separate MPPTs can be applied to each string. This increases the overall efficiency compared to the centralized inverter, and reduces the price, due to mass production.

The AC module is the integration of the inverter and PV module into one electrical device. It removes the mismatch losses between PV modules since there is only one PV module, as well as supports optimal adjustment between the PV module and the inverter and, hence, the individual MPPT. It includes the possibility of an easy enlarging of the system, due to the modular structure. The opportunity to become a "plug-and-play" device, which can be used by persons without any knowledge of electrical installations, is also an inherent feature. On the other hand, the necessary high voltage-amplification may reduce the overall efficiency and increase the price per watt, because of more complex circuit topologies. On the other hand, the AC module is intended to be mass produced, which leads to low manufacturing cost and low retail prices. The present solutions use self-commutated DC-AC inverters, by means of IGBTs or MOSFETs, involving high power quality in compliance with the standards.

3. The Future—Multi—String Inverters, AC Modules and AC Cells

The multi-string inverter is the further development of the string inverter, where several strings are interfaced with their own DC-DC converter to a common DC-AC inverter. This is beneficial. Compared with the centralized system, since every string can be controlled individually. Thus, the operator may start his/her own PV power plant with a few modules. Further enlargements are easily achieved since a new string with DC-DC converter can be plugged into the existing platform. A flexible design with high efficiency is hereby achieved. Finally, the AC cell inverter system is the case where one large PV cell is connected to a DC-AC inverter. The main challenge for the designers is to develop an inverter that can amplify the very low voltage, 0. 5～1. 0 V and 100 W per square meter, up to an appropriate level for the grid, and at the same time reach a high efficiency. For the same reason, entirely new converter concepts are required.

This review has covered some of the standards that inverters for PV and grid applications must fulfill, which focus on power quality, injection of dc currents into the grid, detection of islanding operation, and system grounding. The demands stated by the PV modules have also been reviewed; in particular, the role of power decoupling between the modules and the grid has been investigated. An important result is that the amplitude of the ripple across a PV module should not exceed 3. 0 V in order to have a utilization efficiency of 98% at full generation. Finally, the basic demands defined by the operator have also been addressed, such as low cost, high efficiency, and long lifetime.

The next part of the review was a historical summary of the solutions used in the past, where large areas of PV modules were connected to the grid by means of centralized inverters. This included many shortcomings for which reason the string inverters emerged. A natural development was to add more strings, each with an individual DC-DC converter and MPPT, to the common DC-AC inverter, thus, the multi-string inverters were brought to light. This is believed to be one of the solutions for the future.

Another trend seen in this field is the development of the ac module, where each PV module is interfaced to the grid with its own DC-AC inverter. The historical review was followed with a classification of the inverters: number of power processing stages, type of power decoupling between the PV module and the grid, transformers and types of interconnections between the stage, and types of grid interfaces.

Words and Phases

centralized inverters 集中式逆变器
PV modules 光伏模块
series connections 串联
amplification *n.* 放大
harmonics *n.* 谐波
string inverters 组串式逆变器
semiconductors 半导体
line-frequency *n.* 工频
modular *n.* 模块化
plug-and-play *adj.* 即插即用
self-commutated *adj.* 自（动）换向
compliance *n.* 遵从
hereby *adv.* 据此
islanding operation 孤岛运行
system grounding 系统接地
power decoupling 功率耦合
utilization efficiency 利用效率

Notes

1. The main challenge for the designers is to develop an inverter that can amplify the very low voltage, 0.5～1.0 V and 100 W per square meter, up to an appropriate level for the grid, and at the same time reach a high efficiency.

设计人员面临的主要挑战是研发一种逆变器，可以放大每平方米光伏板上电压为0.5～1.0 V和功率为100 W的极低电压，满足电网并网电压要求同时使得逆变器转换效率达到最高。

2. An important result is that the amplitude of the ripple across a PV module should not exceed 3.0 V in order to have a utilization efficiency of 98% at full generation.

一个重要的结果是PV模块上的纹波幅度不应超过3.0 V，以便在满功率发电时具有98%的利用效率。

3. The historical review was followed with a classification of the inverters: number of power processing stages, type of power decoupling between the PV module and the grid, transformers and types of interconnections between the stage, and types of grid interfaces.

此综述在介绍完逆变器历史发展总结之后，便开始介绍逆变器的分类：功率级数，PV模块和电网之间的功率解耦类型、变压器和不同级之间连接方式、并网连接方式。

Exercises

1. Answer the following questions according to the text

(1) What are the defects of centralized Inverters?

(2) What is themaximum power point tracking (MPPT)?

(3) Compared with centralized inverters, what are the advantages of string inverters?

(4) What is the multi-string Inverters?

(5) What are your views on the future development of photovoltaic (光伏) Inverters?

2. Translate the following sentences into Chinese according to the text

(1) This centralized inverter includes some severe limitations, such as high-voltage DC cables between the PV modules and the inverter, power losses due to a centralized MPPT, mismatch losses between the PV modules, losses in the string diodes, and a nonflexible design where the benefits of mass production could not be reached.

(2) The large amount of harmonics was the occasion of new inverter topologies and system layouts, in order to cope with the emerging standards which also covered power quality.

(3) The opportunity to become a "plug-and-play" device, which can be used by persons without any knowledge of electrical installations, is also an inherent feature.

(4) This review has covered some of the standards that inverters for PV and grid applications must fulfill, which focus on power quality, injection of DC currents into the grid, detection of islanding operation, and system grounding.

(5) A natural development was to add more strings, each with an individual DC-DC converter and MPPT, to the common DC-AC inverter, thus, the multi-string inverters were brought to light. This is believed to be one of the solutions for the future.

3. Translate the following into Chinese

The AC module is the integration of the inverter and PV module into one electrical device. It removes the mismatch losses between PV modules since there is only one PV module, as well as supports optimal adjustment between the PV module and the inverter and, hence, the individual MPPT. It includes the possibility of an easy enlarging of the system, due to the modular structure. The opportunity to become a "plug-and-play" device, which can be used by persons without any knowledge of electrical installations, is also an inherent feature. On the other hand, the necessary high voltage-amplification may reduce the overall efficiency and increase the price per watt, because of more complex circuit topologies. On the other hand, the AC module is intended to be mass produced, which leads to low manufacturing cost and low retail prices. The present solutions use self-commutated DC-AC inverters, by means of IGBTs or MOSFETs, involving high power quality in compliance with the standards.

Unit 6 AC-AC Converter

Text A Review of Three-phase AC-AC Converter Topologies

1. AC-AC Converters with DC-link

Converter systems with either a voltage or a current DC-link are mainly used nowadays for power conversion from a three-phase mains system to a three-phase load with an arbitrary voltage amplitude and frequency, as required, for example, for variable-speed drives. In the case of a converter with a voltage DC-link, the mains coupling can, in the simplest case, be implemented by a diode bridge. A pulse-controlled braking resistor must be placed across the DC-link, or an antiparallel thyristor bridge must be inserted on the mains side to enable generator (braking) operation of the load. The disadvantages are the relatively high mains distortion and high reactive power requirements.

A mains-friendly AC-AC converter with bidirectional power flow can be implemented by coupling the DC-link of a PWM rectifier and a PWM inverter. The DC-link quantity is then impressed by an energy storage element that is common to both stages: a capacitor C_{DC} for the voltage DC-link back-to-back converter [Voltage DC-link back-to-back converter (V-BBC), Fig. 3. 10 (a)] or an inductor L_{DC} for the current DC-link back-to-back converter [current DC-link back-to-back converter (C-BBC), Fig. 3. 10 (b)]. The PWM rectifier is controlled in such a manner that a sinusoidal mains current is drawn, which is in-phase or antiphase with the corresponding mains line voltage. The implementation of the V-BBC and C-BBC requires 12 transistors (typically IGBTs) and 12 diodes or 12 reverse conduction IGBTs (RC-IGBTs) for the V-BBC and 12 reverse blocking IGBTs (RB-IGBTs) for the C-BBC.

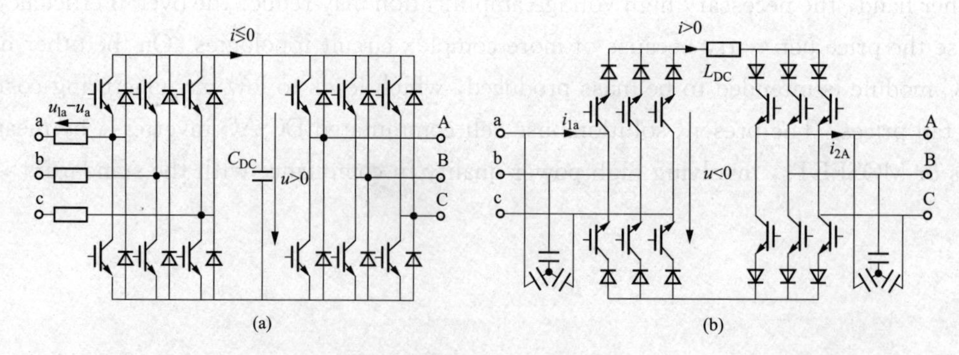

Fig. 3. 10 Basic three-phase AC-AC converter topologies with DC-link energy storage

(a) V-BBC; (b) C-BBC

Due to the DC-link energy storage element, there is an advantage that both converter stages are, to a large extent, decoupled regarding their control for a typical sizing of the energy storage. Furthermore, a constant mains-independent input quantity exists for the PWM inverter stage, which results in a high utilization of the converter's power capability. On the other hand, the DC-link energy storage element can have a relatively large physical volume compared with the total converter volume, and when electrolytic capacitors are used for the DC-link of the V-BBC, the service lifetime of the converter can potentially be reduced.

2. AC-AC Converters without DC-link: Matrix Converter

Aiming for high power densities, it is hence obvious to consider the so-called matrix converter (MC) concepts that enable three-phase AC-AC conversion without any intermediate energy storage element. The physical basis of these systems is the constant instantaneous power provided by a symmetrical three-phase voltage-current system. Conventional (direct) MCs [CMCs; Fig. 3.11 (a)] perform the voltage and current conversions in a single stage. Alternatively, the option of an indirect conversion by means of an indirect MC [IMC; Fig. 3.11 (b)] exists. The IMC requires separate stages for the voltage and current conversions, in a similar way to the V-BBC and the C-BBC, but without an energy storage element in the intermediate link. The implementation of both MC topologies requires 18 IGBTs and 18 diodes, in the basic configuration, or 18 RB-IGBTs for the CMC or 12 RB-IGBTs and 6 RC-IGBTs for the IMC. Thus, the intermediate energy storage element is eliminated at the expense of more semiconductors.

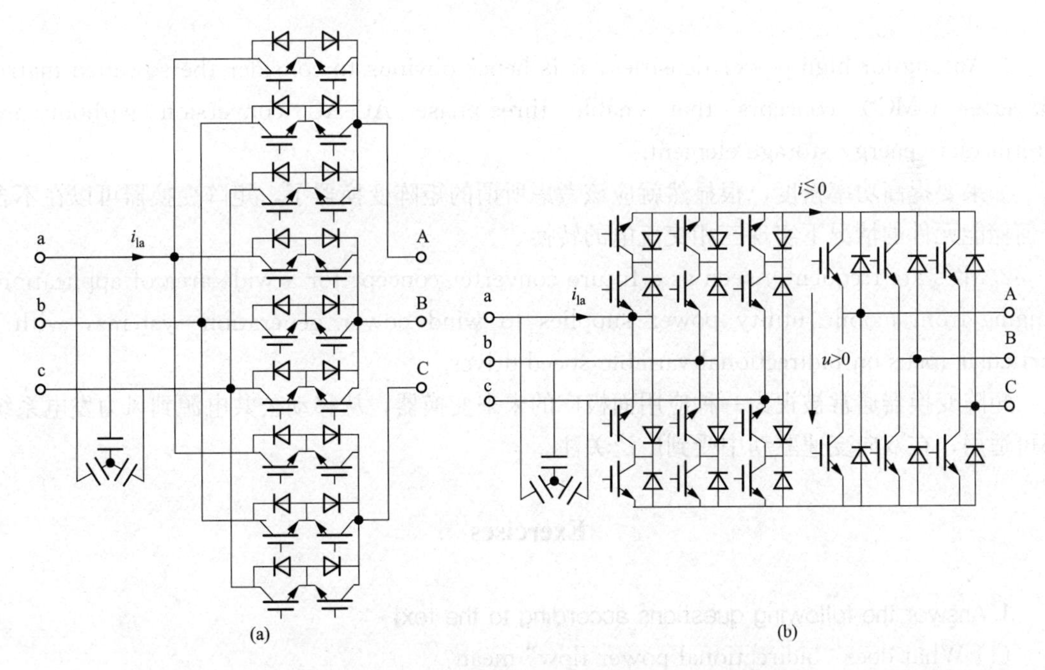

Fig. 3.11 Basic three-phase AC-AC MC topologies
(a) CMC; (b) IMC

MCs are frequently seen as a future converter concept for a wide area of applications ranging from mobile utility power supplies to wind power generation systems, with a particular focus on bidirectional variable-speed drives. Despite intensive research over the last three decades, they have, until now, only achieved low market penetration. Excluding the technical aspects, the reasons for this could be the more complex modulation and dimensioning compared to converters with a DC-link and the high topological variations. The main forced commutated AC-AC converters presented in literature are divided into three subcategories: converters with DC-link energy storage, MCs, and intermediate category of hybrid MCs. A forced commutated AC-AC converter is considered as a MC if it does not require an intermediate energy storage in the power circuit as an essential functional element.

Words and Phases

braking resistor 制动电阻
in-phase, antiphase 同相，反相
electrolytic capacitor 电解电容
Matrix Converter, MC 矩阵变换器
back-to-back converter 背对背变换器
utilization *n.* 利用
dimensioning *n.* 标注尺寸
subcategory *n.* 子分类
Reverse Blocking 反向阻断
Reverse conduction 反向续流
forced commutated 强迫换相

Notes

1. Aiming for high power densities, it is hence obvious to consider the so-called matrix converter (MC) concepts that enable three-phase AC-AC conversion without any intermediate energy storage element.

如果要提高功率密度，很显然就应该考虑所谓的矩阵变换器了，矩阵变换器可以在不含任何储能元件的情况下实现三相交流电的转换。

2. MCs are frequently seen as a future converter concept for a wide area of applications ranging from mobile utility power supplies to wind power generation systems, with a particular focus on bidirectional variable-speed drives.

矩阵变换器通常被视为一种应用面极广的未来变换器，从移动公共电源到风力发电系统都可适用，在双向变速驱动中受到广泛关注。

Exercises

1. Answer the following questions according to the text

(1) What does "bidirectional power flow" mean?

(2) The advantage and disadvantage of DC-link energy storage element?

(3) the advantage and disadvantage of MC?

2. Translate the following sentences into Chinese according to the text

(1) A pulse-controlled braking resistor must be placed across the DC-link, or an

antiparallel thyristor bridge must be inserted on the mains side to enable generator (braking) operation of the load.

(2) A mains-friendly AC-AC converter with bidirectional power flow can be implemented by coupling the DC-link of a PWM rectifier and a PWM inverter.

(3) A constant mains-independent input quantity exists for the PWM inverter stage, which results in a high utilization of the converter's power capability.

(4) Conventional (direct) MCs [CMCs; Fig. 3. 11 (a)] perform the voltage and current conversions in a single stage. Alternatively, the option of an indirect conversion by means of an indirect MC [IMC; Fig. 3. 11 (b)] exists.

(5) A forced commutated AC-AC converter is considered as a MC if it does not require an intermediate energy storage in the power circuit as an essential functional element.

3. Translate the following into Chinese

The PWM output stage (inverter) of the V-BBC, is made up of three bridge legs. Each exhibits the function of a switch that connects the output to either the positive or the negative DC-bus p and n. The switching state of the inverter is defined by (xxx) where x is either p or n; for example, (pnn) means that the output A is connected to p and that the outputs B and C are connected to n.

The PWM input stage (rectifier) of the C-BBC, features the basic functionality of a diode bridge with respect to the conducting state of the power transistors. The control of the power transistors must guarantee that a path for the impressed DC-link current is always available when no separate freewheeling diode is used across the DC-link. Therefore, at least one transistor of the positive and one transistor of the negative bridge-halves must always be held in the ON-state.

Text B Matrix Converter Application

1. Introduction

The MC has primarily been considered for bidirectional variable frequency AC drives for low and later also for medium voltage applications. The drive system integration capability of MCs was demonstrated in 2000 by Klumpner et. al. by attaching a CMC directly to a motor, leading to the so-called Matrix Converter Motor (MCM).

MCs have been widely analyzed for aircraft applications motivated by their potential for a light weight and compact implementation and are still considered as an alternative converter concept for this application area. More recent investigations suggested the MC also for wind generation systems, deep-sea robots, contact-less energy transmission, or AC utility power units. The three-phase to single-phase MC has been recently proposed for distributed energy generator systems, whereas the single-phase MC topology is being mainly investigated as an active front-end for single-phase AC traction applications.

2. AC Motor Drive

The low voltage transfer ratio is often seen as the biggest disadvantage of a matrix converter for an industrial drive is required. Some attempts to address the problem on an overmodulation basis have been performed, but inevitably, input power quality is sacrificed in favor of output drive capability. Work based on minor topology changes, particularly using the indirect matrix converter, has been proposed in at the cost of increased complexity and size. In applications where the load motor in the drive system can be specified and appropriately selected, the voltage transfer ratio limitation is not an issue.

In motor drive where the converter is integrated with or sold with the motor, clearly, the matrix converter should have a size and a weight advantage over competing VSI technologies. A 4kW integrated matrix converter-induction machine drive was applied. The design and construction of a 30kW version was further described by M. Apap. The potential size and weight advantages of the matrix converter and the elevated temperature capability due to the lack of DC-link components lend themselves to aircraft applications.

Several prototype aircraft actuator projects with MC have been reported. P. Wheeler, collaboration with Smiths Aerospace led to the creation of a 7kW matrix converter used to drive a 10000r/min PMSM integrated into an electrohydrostatic actuator. The matrix converter was chosen in this application because of the ability to be driven from a frequency wild supply. This prototype was based on the Infineon Economac matrix converter module. A higher power direct drive EMA was described. Collaboration with Smiths Aerospace, which later became GE Aviation, resulted in the development of both a 20kW integrated matrix converter and a 20kW 10000r/min PMSM to create a fully integrated rudder actuator.

An indirect matrix converter drive was developed in collaboration with MOOG for an EMA application. The same requirements for a variable frequency supply with aircraft power quality specifications were desired as per the previous two examples. The main difference in this project was the requirement to prevent the regeneration of energy back to the supply. This process can be more easily achieved using an indirect matrix converter using a suitable dissipation circuit connected to the standard protective clamp circuit. A comprehensive comparison of the conventional matrix converter, the indirect matrix converter, and the back-to-back PWM rectifier-inverter is described by J. Kola.

A deep sea remotely operated vehicle (ROV) matrix converter drive application was the subject. The extreme pressure experienced by ROVs and the lack of large and fragile DC-link components were the reason that the matrix converter was chosen as a potential topology for the application. Research into the effects of high atmospheric pressure on the constituent parts of typical drive systems was carried out at 300 bar. The use of observer-based sensorless control of a PMSM using the matrix converterare also investigated.

The matrix converter has also been applied to drive the rotor circuit of a doubly fed induction generator for wind turbine applications using direct and indirect matrix converters. This technique has the advantage that a relatively low power four-quadrant power converter can be used to control a high-power generator system. Research into the stability of such systems is presented, and the effects of rotor-side harmonics in a similar system were presented.

A reduced matrix converter (three phase to two phase) was used to control a wind turbine generator and drive a single phase transformer which was then connected through an ac rectifier to a dc transmission line. The efficiency under different modulation and control techniques was addressed for this type of matrix converter. Further work into a novel matrix converter topology to allow the coupling of energy generation resources and the grid was recently presented.

Further industrial interest is outlined, which highlights the characteristics of the matrix converter from a manufacturer' s perspective in terms of cost, competitiveness, and size. The overmodulation performance of the matrix converter is also evaluated for applications where a direct replacement of an industrial VSI drive is required.

3. Wind generation systems

Matrix converters are finding application in the power supply generation area. Instead of the typical motor drive application, an output filter is used in order to provide a voltage source of the desired amplitude and frequency. This concept allows fixed voltage and frequency power supplies to be implemented and driven from variable frequency diesel generators. The issues regarding the control of the output voltage and frequency when using a resonant LC output filter under stringent power quality requirements are described. The operation of the generator at the optimum speed, particularly under lightly loaded conditions, can offer increased fuel efficiency. The application of the matrix converter in polyphase generator systems has been discussed. In this case, the matrix converter transforms not only the input frequency and voltage but also the number of phases. Since the matrix converter circuit is modular, any number of input and output phases can be implemented. Protection strategies for the matrix converter when used as a grid supply converter are discussed.

Words and Phases

bidirectional *adj*. 双向
aircraft applications 航空应用
deep-sea robots 深海机器人
contact-less energy transmission 无线传能
electrohydrostatic actuator 电静液压制动器
collaboration *adj*. 合作
rudder actuator 舵机
clamp circuit 钳位电路
remotely operated vehicle 遥控潜水器
diesel generators 柴油发电机
resonant *n*. 谐振
polyphase generator 多相发电机

Note

1. In motor drive where the converter is integrated with or sold with the motor, clearly, the matrix converter should have a size and a weight advantage over competing VSI

technologies.

若驱动器和电机集成在一起捆绑出售时，矩阵变换器相较于现有的电压型逆变器技术(voltage source inverter，VSI)在体积和重量上有着明显优势。

2. Matrix converters are finding application in the power supply generation area. Instead of the typical motor drive application，an output filter is used in order to provide a voltage source of the desired amplitude and frequency.

矩阵变换器正在发电领域得到应用。与电机驱动应用不同，通过增加一个输出滤波器来得到想要的幅值和频率。

Exercises

1. Answer the following questions according to the text

(1) Why MCs have been widely analyzed for aircraft applications?

(2) The disadvantages of overmodulation with MC?

(3) What are the differences between the two matrix converter applied to the aero-electrostatic actuator，EHA?

(4) How matrix converters connect power supply generation and grids?

2. Translate the following sentences into Chinese according to the text

(1) The MC has primarily been considered for bidirectional variable frequency AC drives for low and later also for medium voltage applications.

(2) MCs have been widely analyzed for aircraft applications motivated by their potential for a light weight and compact implementation and are still considered as an alternative converter concept for this application area.

(3) In applications where the load motor in the drive system can be specified and appropriately selected，the voltage transfer ratio limitation is not an issue.

(4) This process can be more easily achieved using an indirect matrix converter using a suitable dissipation circuit connected to the standard protective clamp circuit.

(5) Matrix converters are finding application in the power supply generation area. Instead of the typical motor drive application，an output filter is used in order to provide a voltage source of the desired amplitude and frequency.

3. Translate the following into Chinese.

The matrix converter offers many potential benefits to the power converter industry. It will not be the best solution for all uses，but it offers significant advantages for many different applications. While，for many years，it seemed that the matrix converter would be restricted to a small range of niche areas，the commitment to invest in matrix converters from several large industrial drives manufacturers may see the start of an industry wide uptake of this technology.

Unit 7 Design Considerations of Practical Converter

Text A Supplementary Components of Power Converters

A practical power electronic converter is a complex systems comprising several subsystems and numerous components. Many of them are not shown in the converter circuit diagrams, which are usually limited to the power circuit and, sometimes, a block diagram of the control system. The supplementary components for modern power electronic converters include:

Drivers for individual semiconductor power switches, which provide the switching signals, interfacing the switches with the control system.

Protection circuits, which safeguard converter switches and sensitive loads from excessive currents, voltages, and temperatures.

Snubbers, which protect switches from transient overvoltages and overcurrents at turn-on and turn-off and reduce the switching losses.

1. Drivers

Depending on the type of switches, converter topology, and voltage levels, various driver configurations are employed in power electronic converters. A driver, activated by a logic-level signal from the control system, must be able to provide a sufficiently high voltage or current to the controlling electrode, gate or base, to cause an immediate turn-on. The on-state of the switch must then be safely maintained until turn-off.

As seen form Fig. 3. 12, the driver must ensure electrical isolation between the low-voltage control system and high-voltage power circuit. It is realized using pulse transformers (PTRs) or optical coupling. The latter is performed by placing a light-emitting diode (LED)

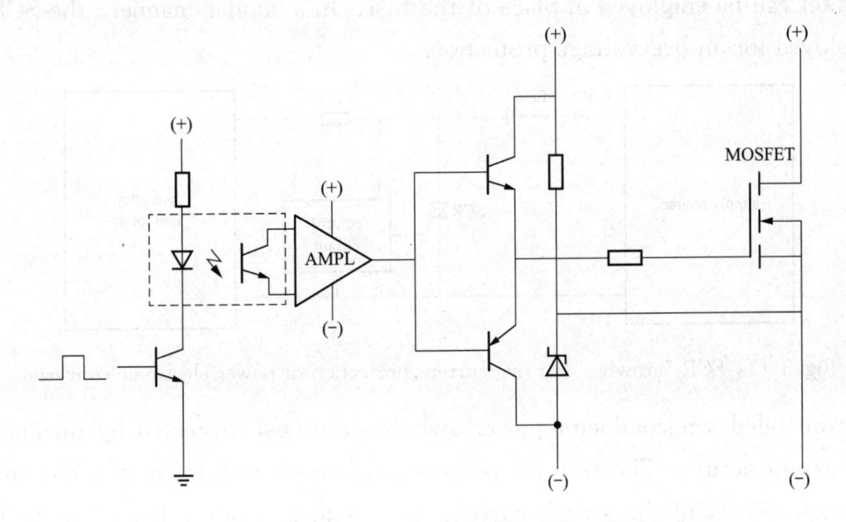

Fig. 3. 12 Driver for a power MOSFET with optical isolation

in the vicinity of a light-activated semiconductor device. Alternately, instead of transferring light signal through free space, a fiber-optic cable can be employed. Because of the fundamentally different driving signal requirements, different solutions are used for semi-controlled thyristors (SCRs, TRIACs, BCTs), current-controlled switches (GTOs, IGCTs, power BJTs), and voltage-controlled hybrid devices (power MOSFETs and IGBTs).

2. Overcurrent protection schemes

Semiconductor power switches can easily suffer a permanent damage if a short circuit occurs somewhere in the converter or the load, or if an overcurrent occurs due to an excessive load, that is, a too low load impedance and/or counter EMF. The following three basic approaches to overcurrent protection are used in power electronic converters: ① fuses; ② an SCR "crowbar" arrangement; ③turning the switches off when overcurrent is detected.

High-power slow-reacting semiconductor power devices, such as diodes, SCRs, or GTOs, are protected by special, fast-melting fuses connected in series with each device. The fuses are made of thin silver bands, not thicker than a tenth of an inch, usually with a number of rows of punched holes. One or more of such bands are packed in sand and enclosed in a cylindrical ceramic body with tinned mounting brackets. The fuse coordination I^2t parameter of a fuse must be less than that of the protected device, but not so low as to cause a breakdown under regular operating conditions. A properly selected fuse should melt within a half cycle of the 60 Hz (or 50 Hz) voltage.

Low-cost power electronic converters can be protected by a single fuse between the supply source and the input to the converter. A more sophisticated solution, illustrated in Fig. 3. 13, involves an SCR "crowbar", connected across the input terminals of a converter. The input current to the converter is monitored by sensing the voltage drop across the low-resistance resistor R or employing a current sensor. If overcurrent is detected, the SCR is fired shorting the supply source and causing a meltdown of the input fuse. Alternately, a fast circuit breaker can be employed in place of the fuse. In a similar manner, the SCR crowbar can be employed for an overvoltage protection.

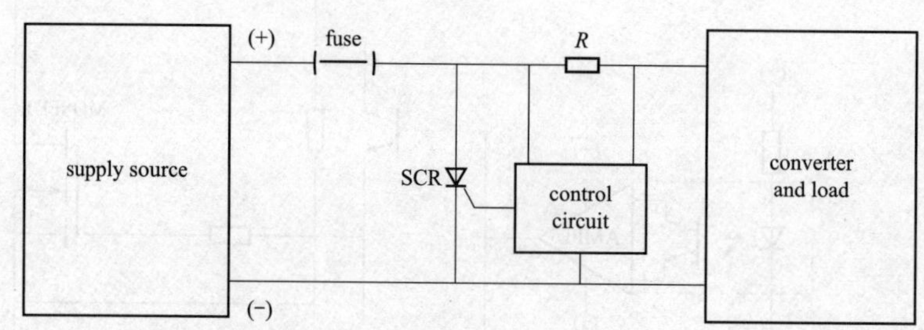

Fig. 3. 13 SCR "crowbar" for overcurrent protection of power electronic converter

Fully controlled semiconductor power switches are best protected by turning them off when overcurrent occurs. The turn-off process is often slowed down to avoid an excessive rate of change, di/dt, of the switch current that could generate a hazardous voltage spike across the device. As exemplified by the system in Fig. 3. 13, dedicated overcurrent protection circuits are often incorporated in the modern drivers.

3. Snubbers

Of all the possible modes of operation of semiconductor devices, switching between two extreme states is the most trying one, subjecting switches in power electronic converters to various stresses. For example, if no measures were taken, a rapid change of the switched current at turn-off would produce potentially damaging voltages spikes in stray inductances of the power circuit. At turn-on, a simultaneous occurrence of high voltage and current could take the operating point of a switch well beyond the safe operating area (SOA). Therefore, switching-aid circuits called snubbers must often accompany semiconductor power switches. Their purpose is to prevent transient overvoltages and overcurrents, attenuate excessive rates of changes of voltage and current, reduce switching losses, and ensure that the switch does not operate outside its SOA. Snubbers also help to maintain uniform distribution of voltages across the switches that are connected in series to increase the effective voltage rating or currents in the switches that are connected in parallel to increase the effective current rating. Functions and configurations of snubbers depend on the type of switch and converter topology.

Words and Phases

supplementary *adj*. 外加的
driver *n*. 驱动
snubber circuit 缓冲器，缓冲电路，吸收电路
electrode *n*. 电极
gate *n*. 栅极
base *n*. 基极
electrical isolation 电气隔离
pulse transformer 脉冲变压器
optical coupling 光耦合
light-emitting diode 发光二极管
fiber-optic cable 光缆
semi-controlled thyristor 半控型晶闸管
permanent damage 永久性损坏
crowbar *n*. 撬棒电路，消弧电路
ceramic *n*. 陶瓷
breakdown *n*. 击穿
voltage spike 电压尖峰
stray inductance 杂散电感
safe operating area 安全工作区
switching loss 开关损耗

Note

1. A driver, activated by a logic-level signal from the control system, must be able to provide a sufficiently high voltage or current to the controlling electrode, gate or base, to cause an immediate turn-on.

驱动器接收来自控制系统的逻辑信号，必须能够为被控电极（栅极或基极）提供足够大的电压或电流以实现快速开通。

2. Semiconductor power switches can easily suffer a permanent damage if a short circuit occurs somewhere in the converter or the load, or if an overcurrent occurs due to an excessive

load, that is, a too low load impedance and/or counter-EMF.

半导体开关器件在以下两种情况下极易造成永久性损坏，当变换器或负载发生短路时，或当过负荷（过小的负载阻抗或者反电动势）引发过电流时。

3. Of all the possible modes of operation of semiconductor devices, switching between two extreme states is the most trying one, subjecting switches in power electronic converters to various stresses.

在半导体器件所有可能的操作模式中，在两个极端状态下的开关是最难应对的，这将会使电力电子变换器中的开关器件承受多种应力。

Exercises

1. Answer the following questions according to the text

(1) How to provide electrical isolation between the low-voltage control system and high-voltage power circuit?

(2) How to choose the parameter of the fuse for high-power slow-reacting semiconductor power devices?

(3) What is the working principle of the SCR "crowbar"?

(4) Why are snubbers of great importance for semiconductor power switches?

2. Translate the following sentences into Chinese according to the text

(1) The latter is performed by placing a light-emitting diode (LED) in the vicinity of a light-activated semiconductor device.

(2) Because of the fundamentally different driving signal requirements, different solutions are used for semi-controlled thyristors (SCRs, TRIACs, BCTs), current-controlled switches (GTOs, IGCTs, power BJTs), and voltage-controlled hybrid devices (power MOSFETs and IGBTs).

(3) High-power slow-reacting semiconductor power devices, such as diodes, SCRs, or GTOs, are protected by special, fast-melting fuses connected in series with each device.

(4) The fuse coordination I^2t parameter of a fuse must be less than that of the protected device, but not so low as to cause a breakdown under regular operating conditions.

(5) The turn-off process is often slowed down to avoid an excessive rate of change, di/dt, of the switch current that could generate a hazardous voltage spike across the device.

3. Translate the following into Chinese

Of all the possible modes of operation of semiconductor devices, switching between two extreme states is the most trying one, subjecting switches in power electronic converters to various stresses. For example, if no measures were taken, a rapid change of the switched current at turn-off would produce potentially damaging voltages spikes in stray inductances of the power circuit. At turn-on, a simultaneous occurrence of high voltage and current could take the operating point of a switch well beyond the safe operating area (SOA). Therefore, switching-aid circuits called snubbers must often accompany semiconductor power switches.

Their purpose is to prevent transient overvoltages and overcurrents, attenuate excessive rates of changes of voltage and current, reduce switching losses, and ensure that the switch does not operate outside its SOA. Snubbers also help to maintain uniform distribution of voltages across the switches that are connected in series to increase the effective voltage rating or currents in the switches that are connected in parallel to increase the effective current rating. Functions and configurations of snubbers depend on the type of switch and converter topology.

Text B Supplementary Systems of Power Converters

1. Control

To function efficiently and safely, a power electronic converter must be properly controlled, that is, the process of power conversion must be accompanied by concurrent information processing. Various control technologies have been employed over the several last decades, starting with the analog electronic circuits with discrete components and progressing to the contemporary integrated microelectronic digital systems. It is worth mentioning that simple analog control is still employed in many low-power DC-DC converters, especially those with very high switching frequencies.

In practical applications, a power electronic converter usually constitutes a part of a larger engineering system, such as an adjustable speed drive or active power filter. The control system of the converter is then subordinated to a master controller, for instance, that of speed of a motor fed from the converter. The main task of the converter control system is to generate signals for semiconductor power switches that result in the desired fundamental output voltage or current. Other "housekeeping chores" are also performed, for example, control of electromechanical circuit breakers connecting the converter with the supply system. Often, especially in expensive, high-power converters, the control system also monitors operating conditions. When a failure is detected, the system turns off the converter and displays the diagnosis. Generally, the system draws information from the human operator, sensors, and master controllers and converts it into switching signals for the converter switches and external circuit breakers. The control system also supplies information about operation of the converter back to the operator, via displays, indicators, or recorders.

As an example of a digital control system, a block diagram of an adjustable-speed AC drive governed by a digital control system based on a DSP controller is illustrated in Fig. 3. 14. The adjustable-frequency and adjustable-magnitude currents for the three-phase AC motor are produced by an inverter, whose three phases are independently controlled by current regulators, CRA through CRC. Each current regulator receives a reference signal from the controller, compares it with the signal obtained from current sensors, and generates appropriate switching signals for the inverter. The control system reacts to the speed control signal from a speed sensor. Most signals are digital, so if the inexpensive analog current sensors are employed, analog-to-digital (A/D) converters must be used.

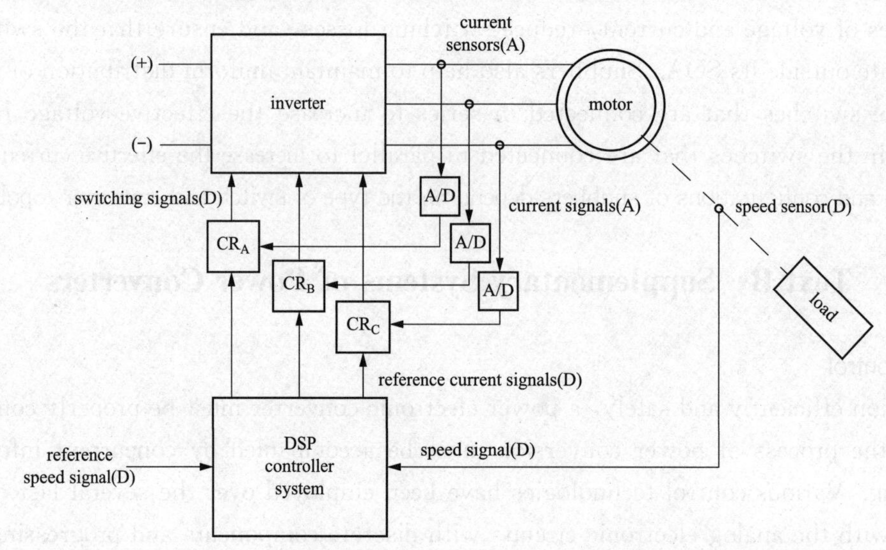

Fig. 3. 14 Block diagram of an adjustable-speed AC drive

(A)—analog; (D)—digital

As an alternative, the currents could be measured by digital sensors, with the current control incorporated in the operating algorithm of the DSP controller. In that case, the input signals to the controller would include those from the current sensors. If the drive system in question is a part of a larger process, the reference speed signal can come from another, hierarchically superior controller.

2. Cooling

Losses in power electronic converters produce heat that must be transferred away and dissipated to the surroundings. The major losses occur in semiconductor power switches, while their small size limits their thermal capacity. High temperatures of the semiconductor structures cause degradation of electrical characteristics of switches, such as the maximum blocked voltage or the turn-off time. Serious overheating can lead to destruction of a semiconductor device in a short time. To maintain a safe temperature, a power switch must be equipped with a heat sink (radiator) and be subjected to at least the natural convection cooling. The heat generated in the switch is transferred via the heat sink to the ambient air, which then tends to move upward and away from the switch.

Forced air cooling, very common in practice, is more effective than the natural cooling. The cooling air is propelled by a fan, typically placed at the bottom of the cabinet that houses the power electronic converter. Slotted openings at the top part of the cabinet allow the heated air to escape to the surroundings. Energy consumption of the low-power fan does not tangibly affect overall efficiency of the converter.

If power density of a converter, that is, the rated power-to-weight ratio, is very high, liquid cooling may be needed. Water, automotive coolant, or oil can be used as the cooling medium. The fluid is forced through extended hollow copper or aluminum bars to which semiconductor power switches are bolted. Thanks to the high specific heat of water, water cooling is very effective although it poses a danger of corrosion. On the other hand, oil, whose specific heat is less than half of that of water, has much better insulating and protecting properties.

Words and Phases

adjustable speed drive 变速驱动
active power filter 有源电力滤波器
diagnosis *n.* 诊断
current regulator 电流调节器
current sensor 电流传感器
speed sensor 转速传感器
analog-to-digital converter 模拟数字变换器
dissipate *v.* 消散
thermal capacity 热容
blocked voltage 阻断电压
overheat *v.* 过热
forced air cooling 强迫风冷
liquid cooling 液冷

Note

1. Various control technologies have been employed over the several last decades, starting with the analog electronic circuits with discrete components and progressing to the contemporary integrated microelectronic digital systems.

过去几十年已有多种控制技术获得应用，从基于分立元件的模拟电子电路系统发展到现代的集成微电子数字系统。

2. Each current regulator receives a reference signal from the controller, compares it with the signal obtained from current sensors, and generates appropriate switching signals for the inverter.

每个电流调节器从控制器获得一个参考信号，并将该信号与从传感器获得的信号进行对比，进而产生对应的开关信号用于逆变器控制。

3. As an alternative, the currents could be measured by digital sensors, with the current control incorporated in the operating algorithm of the DSP controller.

作为替代，电流还可以通过数字传感器测量，进而可使得电流控制器集成到 DSP 控制器的操作算法中。

Exercises

1. Answer the following questions according to the text

(1) What is the main task of the converter control system?

(2) What is the harm of overheating?

(3) How many cooling techniques are mentioned in the text?

2. Translate the following sentences into Chinese according to the text

(1) In practical applications, a power electronic converter usually constitutes a part of a larger engineering system, such as an adjustable speed drive or active power filter.

(2) Generally, the system draws information from the human operator, sensors, and master controllers and converts it into switching signals for the converter switches and external circuit breakers.

(3) The adjustable-frequency and adjustable-magnitude currents for the three-phase AC motor are produced by an inverter, whose three phases are independently controlled by current regulators, CRA through CRC.

(4) Losses in power electronic converters produce heat that must be transferred away and dissipated to the surroundings.

3. Translate the following into Chinese

Forced air cooling, very common in practice, is more effective than the natural cooling. The cooling air is propelled by a fan, typically placed at the bottom of the cabinet that houses the power electronic converter. Slotted openings at the top part of the cabinet allow the heated air to escape to the surroundings. Energy consumption of the low-power fan does not tangibly affect overall efficiency of the converter.

If power density of a converter, that is, the rated power-to-weight ratio, is very high, liquid cooling may be needed. Water, automotive coolant, or oil can be used as the cooling medium. The fluid is forced through extended hollow copper or aluminum bars to which semiconductor power switches are bolted. Thanks to the high specific heat of water, water cooling is very effective although it poses a danger of corrosion. On the other hand, oil, whose specific heat is less than half of that of water, has much better insulating and protecting properties.

Unit 8 Modulation and Control

Text A Pulse Width Modulation (PWM) Techniques

1. Introduction

In order to vary the speed in AC drives, it is necessary to control both the voltage and the frequency applied to the motor terminals. The most common way is that to use an inverter bridge, that consists of six switches which connect each motor terminal to either the positive or the negative rail of a constant dc voltage source. The basic principle is irrespective of the type of the used power devices. However, the maximum operation frequency is an important factor in determining the modulation strategy to be used. some basic PWM techniques are summarized.

2. Sinusoidal Pulse Width Modulation (SPWM)

The sinusoidal Pulse Width Modulation concept is quite simple and is illustrated by Fig. 3. 15. It can be seen that the pulses in the output waveform have a sine weighting equivalent to the reference waveform. This method was realized first with analog circuits, but, at the present time, digital implementations are preferred. The harmonic analysis of this waveform shows a fundamental component proportional to the amplitude of the reference sinewave. There are also harmonic components at the carrier frequency which are large and uncontrolled. The effect of the carrier frequency harmonic reduces as its frequency increases since the effect is limited largely by the motor reactances (frequency dependent). The best choice of carrier frequency is therefore the highest at which the inverter power devices can operate.

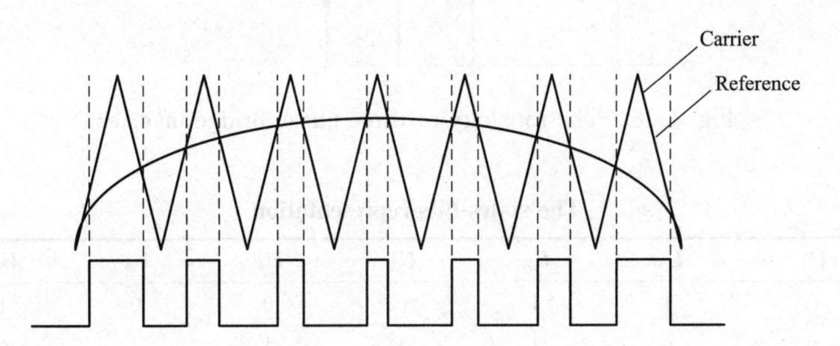

Fig. 3. 15 Pulse Width Modulation using the triangular carrier method with a sinewave reference

The best practical implementation of this scheme is usually by digital system using up-down counters in place of the triangular waveform and lookup tables to determine the reference value at any point.

3. Space Vector Pulse Width Modulation: SVPWM

Since digital AC drives are becoming as an industrial standard, the traditional triangular carrier method has been overcome by the "space vector" representation more suitable for digital implementation. In this paragraph some basic features of this method will be reviewed and the relation with the traditional triangle-sine method will be focused on. Switching on and off each switch of the three inverter legs, three independent voltage waveforms can be generated. Almost generally, motor loads are with insulated neutral or delta connected. Thus, motor senses only the line to line voltages and is insensitive to the common-mode component of the line to neutral voltage. As a consequence, this latter can be varied without any consequence on the load side (there is one degree of freedom). However, the line to line independent voltages are only two, the third one (must sum to zero with the other two) is dependent by the other two.

The three phase inverter is constituted by six switches (as shown Fig. 3. 16) and there are eight possible inverter configurations: six active states and two zero states, In correspondence of each configuration, the six switches have a well defined state: ON or OFF. As a consequence, with three bits, one for each inverter leg (a, b, c), we can identify dl the possible inverter configurations. In fact, we can say that for each leg the bit is 1 when the switch on the top is closed and 0 when the switch on the bottom is closed instead. In Tab. 3. 1 the states-bits representation is shown.

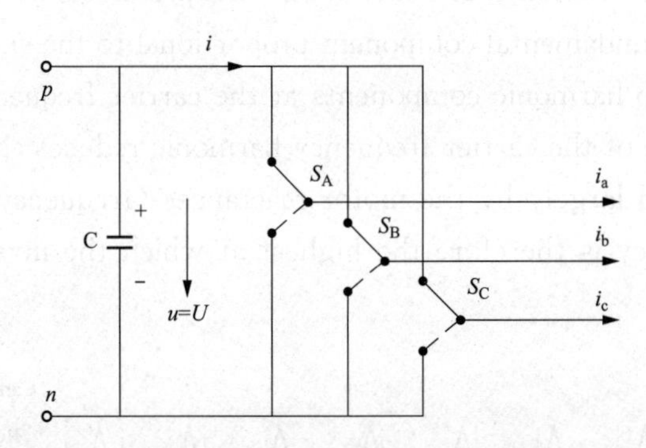

Fig. 3. 16 The topology of three phase bridge inverter

Tab. 3. 1 The states-bits representation

Voltage vector	U_0	U_1	U_2	U_3	U_4	U_5	U_6	U_7
S_A	0	1	1	0	0	0	1	1
S_B	0	0	1	1	1	0	0	1
S_C	0	0	0	0	1	1	1	1

As well known, in a three phase system, the three sinusoidal voltages, 120° apart, can be represented by a rotating vector, whose projections on the fixed three phase axes are, instant by instant, the three sinusoidal waves. As a consequence, the three sinusoidal references can be represented by a voltage reference Vector $\boldsymbol{U}_{\text{ref}}$ Thus, a voltage reference

vector $\boldsymbol{U}_{\mathrm{Ref}}$ can be synthesized only by a combination of these eight states. If the output frequency is much lower than the switching frequency, in a switching period T_s the reference voltage vector can be considered constant. As a consequence, in a "time average" sense, the voltage reference vector $\boldsymbol{U}_{\mathrm{Ref}}$ in a switching period T_s, can be approximated by two non-zero voltage inverter states. En addition, to keep the switching frequency constant, the remainder of the switching period is spent on the zero state.

In Fig. 3. 17 (a) the inverter states are shown and in Fig. 3. 17 (b) the reference voltage vector synthesis is depicted. To obtain the minimum switching frequency of each inverter leg, it is necessary to arrange the switching sequence in such a way that the transition from one state to the next is performed by switching only one inverter leg. In order to match these rules, for example, in the case represented in Fig. 3. 17 (b), the switching sequence has to be: 0-1-2-7-7-2-1-0.

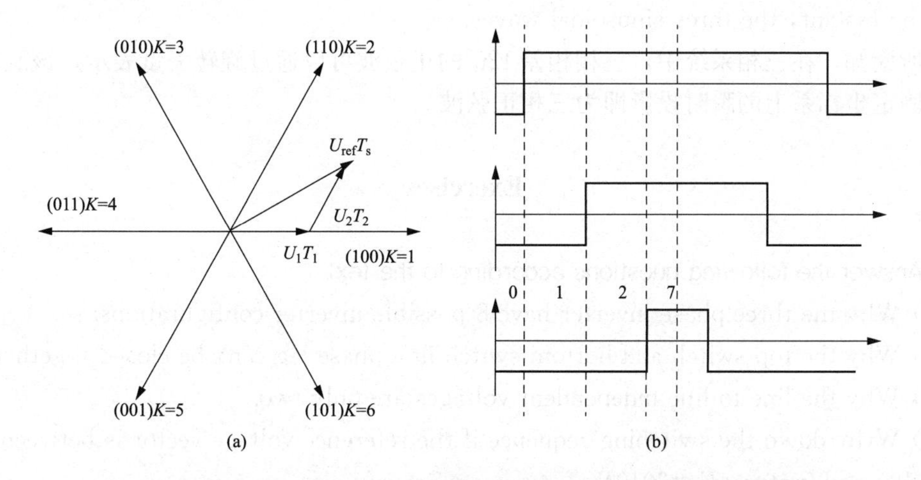

Fig. 3. 17 Principle of SVPWM for three phase bridge inverter

(a) Voltage Vector reference synthesis in a switching period; (b) optimum pulse pattern of svpwm

Words and Phases

irrespective *adv.* 不管，与……无关

Pulse Width Modulation 脉宽调制

fundamental component 基波分量

reactance *n.* 电抗

lookup tables 查表

up-down counter 可逆计数器

triangular waveform 三角波

space vector 空间矢量

triangle-sine method 三角正弦波法

inverter legs 逆变器桥臂

insulated neutral 绝缘中性点

degree of freedom 自由度

rotating vector 旋转矢量

projection *n.* 投影

synthesize *n.* 合成

approximate *adj.* 近似的，估计的

Note

1. The effect of the carrier frequency harmonic reduces as its frequency increases since the effect is limited largely by the motor reactances (frequency dependent).

这些谐波分量频率越高，越受制于电机的电抗效应（与频率相关），所造成的影响也越小。

2. Thus, motor senses only the line to line voltages and is insensitive to the common-mode component of the line to neutral voltage.

因此，电机只对线电压敏感，而对相电压中的共模分量不敏感。

3. As well known, in a three phase system, the three sinusoidal voltages, 120° apart, can be represented by a rotating vector, whose projections on the fixed three phase axes are, instant by instant, the three sinusoidal waves.

众所周知，在三相系统中，三相相差120°的正弦波可以通过旋转矢量表示，该旋转矢量在三相固定坐标系上的瞬时投影即为三相正弦波。

Exercises

1. Answer the following questions according to the text

(1) Why the three phase inverter have 8 possible inverter configurations.

(2) Why the top switch and bottom switch in a phase leg can't be closed together.

(3) Why the line to line independent voltages are only two.

(4) Write down the switching sequence if the reference voltage vector is between vector ($K=2110$) and vector ($K=3010$)

2. Translate the following sentences into Chinese according to the text

(1) The harmonic analysis of this waveform shows a fundamental component proportional to the amplitude of the reference sinewave.

(2) As a consequence, this latter can be varied without any consequence on the load side.

(3) The three phase inverter is constituted by six switches and there are eight possible inverter configurations: six active states and two zero states, In correspondence of each configuration, the six switches have a well defined state: ON or OFF.

(4) To obtain the minimum switching frequency of each inverter leg, it is necessary to arrange the , switching sequence in such a way that the transition from one state to the next is performed by switching only one inverter leg.

3. Translate the following into Chinese

Thus, a voltage reference vector $\boldsymbol{U}_{\text{Ref}}$ can be synthesized only by a combination of these eight states. If the output frequency is much lower than the switching frequency, in a switching period T_s the reference voltage vector can be considered constant. As a consequence, in a "time average" sense, the voltage reference vector U_{ref} in a switching period T_s, can be approximated by two non-zero voltage inverter states. In addition, to keep the switching frequency constant, the remainder of the switching period is spent on the zero state.

Text B Control Method in Power Electronics

1. Introduction

Control strategies for power electronics and drive systems have been a research subject since more than half a century. The preferred concepts in industrial applications are linear control combined with modulation schemes and nonlinear control based on hysteresis bounds. Research on these concepts started using on analogue hardware, which requires a low conceptual complexity. On the other hand, powerful digital hardware (DSP and microprocessor) is now available which can be used also for more complex concepts e.g. direct torque control (DTC) and predictive control.

2. Field-oriented control (FOC) with modulation schemes

Fig. 3. 18 shows a basic block diagram of the field-oriented control (FOC) for an induction motor (IM) supplied drive by Frequency converter. The principle of this control scheme is to mathematically achieve a rotor flux linkage vector λ_r of a motor aligned to the d-axis of a rotating frame (the synchronous frame), as shown in Fig. 3. 19.

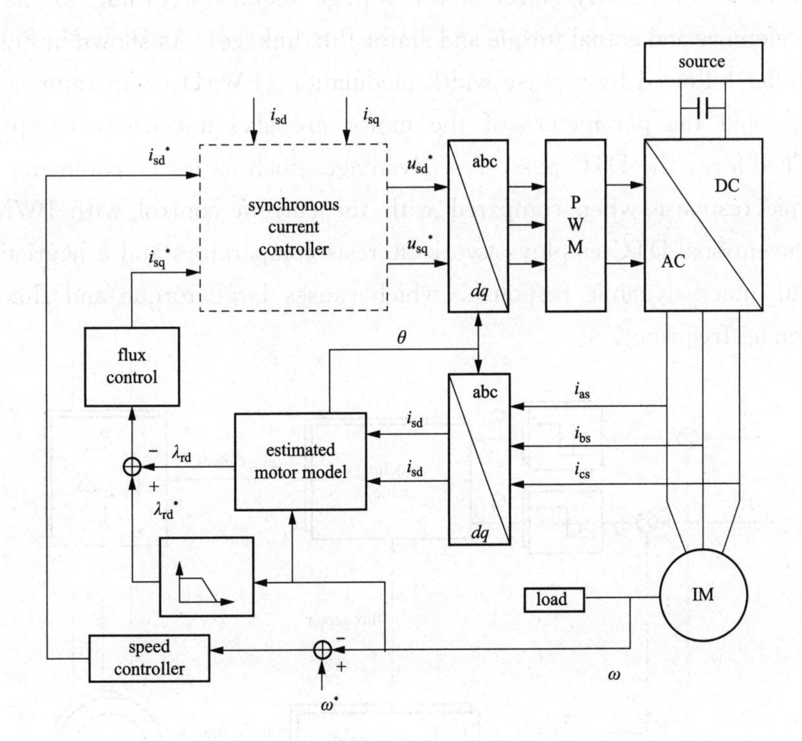

Fig. 3. 18 Basic block diagram for field-oriented control of driving motor IM through a PWM voltage source inverter

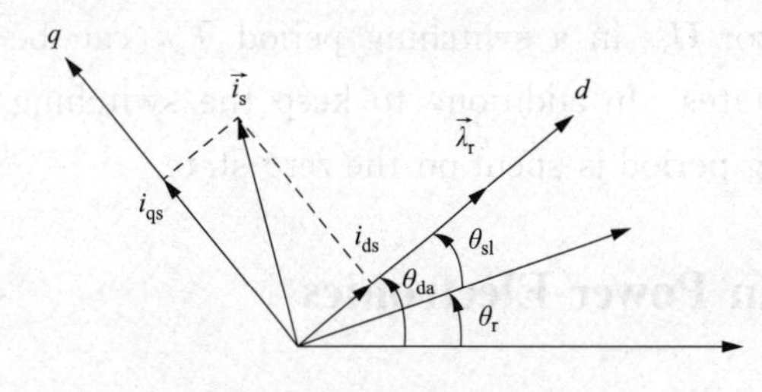

Fig. 3. 19 The frame transformation

The controls scheme is based on synchronous current control and a virtual motor, which numerically estimates the model outputs of the motor in the synchronous frame. The actual synchronous currents i_{sd} and i_{sq} from the three-phase motor are obtained through the reference frame transformation, which requires the displacement angle θ_{da} of the rotor flux linkage along the d-axis.

As shownin Fig. 3. 18, two loops are considered to control the position or mechanical speed of the rotor. The outer loop involves the speed ω_{mech} of the shaft of the motor IM. It generates a q-axis reference current i_{sq}^* through comparison with the reference speed ω_{mech}^* and the speed controller. The synchronous current controller is an inner loop that is used to generate the reference signals of the pulse width modulation (PWM) inverter. This is achieved using two PI controllers and the inverse of the coordinate transformation followed by a PWM modulator that generates pulsing signals to control the DC/AC converter.

3. Direct torque control (DTC) and Model-based predictive control (MPC)

(1) Direct Torque Control (DTC)

Direct torque control (DTC) has become popular in AC motor drives. The basic principle of DTC is to directly select stator voltage vectors according to the differences between the reference and actual torque and stator flux linkage, As shown in Fig. 3. 20. The current controller followed by a pulse width modulation (PWM) comparator is not used in DTC systems, and the parameters of the motor are also not used, except the stator resistance. Therefore, the DTC possesses advantages such as lesser parameter dependence and fast torque response when compared with the current control with PWM schemes. However, conventional DTC employs two hysteresis comparators and a heuristic switching table to obtain quick dynamic response, which causes large torque and flux ripple and variable switching frequency.

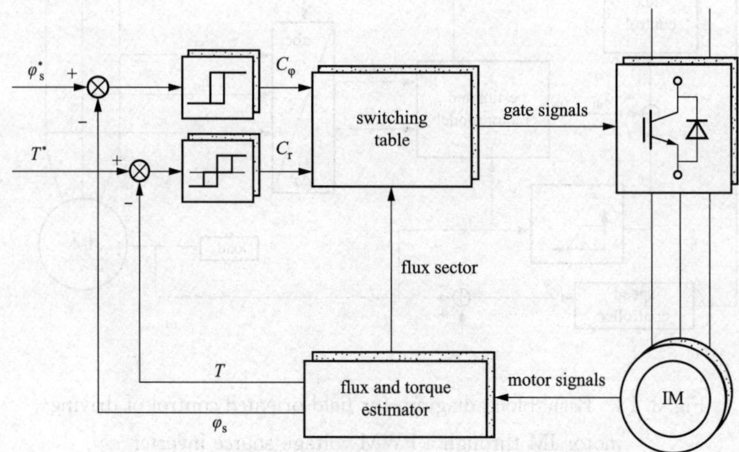

Fig. 3. 20 The basic principle of DTC

In the past decades, numerous schemes have been proposed to address these problems of conventional DTC. Many of them employ space vector modulation (SVM) to produce continuous voltage vectors, which can adjust the torque and flux more accurately and moderately, hence the torque and flux ripples were reduced while obtaining fixed switching frequency. Another merit of using SVM is that the sampling frequency does not need to be as high as that in conventional DTC. Various methods are proposed to obtain the commanding voltage vector, including deadbeat control, sliding mode control, PI controller etc. In the SVM-based DTC schemes, rotary coordinate transformation is often needed, which is more computationally intensive than the conventional DTC.

Another category of modified DTC does not need the SVM block and all the calculations were implemented in stationary coordinate, hence preserving the merits of conventional DTC. Multilevel inverter was introduced to obtain more voltage vectors; however, the hardware cost and system complexity are increased. For two level inverter, more voltage vectors can be synthesized by dividing one sampling period into several intervals, thus a more accurate and complex switching table can be constructed. These methods achieve excellent performance, but they are usually complicated and rely much on the knowledge of motor parameters.

(2) Model-based predictive control (MPC)

Model-based predictive control (MPC) (as shown Fig. 3. 21) for power converters and drives is a control technique that has gained attention in the research community. In Fig. 3. 21, x^* is the reference command, x^* is the predicted value for variable x_i. The converter presnts J different switching states and S_{opt} is the switching state that minimizes the cost function. The main reason for this is that although MPC presents high computational burden, it can easily handle multivariable case and system constraints and nonlinearities in a very intuitive way. Taking advantage of that, MPC has been successfully used for different applications such as an active front end (AFE), power converters connected to resistor-inductor RL loads, uninterruptible power supplies, and high performance drives for induction machines, among others.

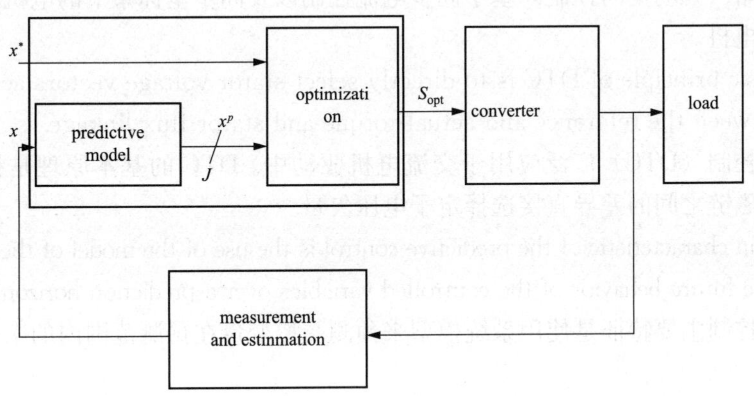

Fig. 3. 21 MPC generic control diagram

The main characteristic of the predictive control is the use of the model of the system for the prediction of the future behavior of the controlled variables over a prediction horizon，N. This information is used by the MPC control strategy to provide the control action sequence for the system by optimizing a user-defined cost function. It should be noted that the algorithm is executed again every sampling period and only the first value of the optimal sequence is applied to the system at instant k. The cost function can have any form，but it is usually defined as

$$g=\sum_{i}\lambda_i(x_i^*-x_i^p)^2 \tag{3-2}$$

Where x_i^* is the reference command，x_i^p is the predicted value for variable x_i，λ_i is a weighting factor，and index i stands for the number of variables to be controlled. In this simple way，it is possible to include several control objectives（multivariable case），constraints，and nonlinearities. The predicted values，x_i^p are calculated by means of the model of the system to be controlled.

Words and Phases

direct torque control 直接转矩控制
predictive control 预测控制
Field-oriented control（FOC） 磁场定向控制
align *v.* 排整齐，校准，使一致
heuristic *adj.* 启发式
moderately *adj.* 适当的，适度的
deadbeat control 有限拍控制
sliding mode control 滑模控制
Model-based predictive control（MPC） 模型预测控制
intuitive *adj.* 直观地，易懂的
active front end 有源前端

Note

1. The controls scheme is based on synchronous current control and a virtual motor，which numerically estimates the model outputs of the motor in the synchronous frame.

该控制策略（磁场定向控制）基于同步电流控制以及同步坐标系下的电机输出观测模型所获得的虚拟电机。

2. The basic principle of DTC is to directly select stator voltage vectors according to the differences between the reference and actual torque and stator flux linkage.

直接转矩控制（DTC）广泛应用于交流电机驱动中。DTC 的基本原理是根据参考和实际转矩与定子磁链之间的差异直接选择定子电压矢量

3. The main characteristic of the predictive control is the use of the model of the system for the prediction of the future behavior of the controlled variables over a prediction horizon，N.

模型预测控制主要特征是使用系统模型来预测受控变量在预测范围内的未来行为 N。

Exercises

1. Answer the following questions according to the text

(1) The advantage of the field-oriented control (FOC) of an induction motor?

(2) The advantage and disadvantage of the direct torque control (DTC)?

(3) The advantage and disadvantage of the SVM-based DTC schemes?

(4) The advantage and disadvantage of the model-based predictive control (MPC)

2. Translate the following sentences into Chinese according to the text.

(1) The preferred concepts in industrial applications are linear control combined with modulation schemes and nonlinear control based on hysteresis bounds.

(2) The principle of this control scheme is to mathematically achieve a rotor flux linkage vector λ_r of a motor aligned to the d-axis of a rotating frame (the synchronous frame).

(3) Direct torque control (DTC) has become popular in AC motor drives. The basic principle of DTC is to directly select stator voltage vectors according to the differences between the reference and actual torque and stator flux linkage.

(4) Many of them employ space vector modulation (SVM) to produce continuous voltage vectors, which can adjust the torque and flux more accurately and moderately, hence the torque and flux ripples were reduced while obtaining fixed switching frequency.

(5) The main reason for this is that although MPC presents high computational burden, it can easily handle multivariable case and system constraints and nonlinearities in a very intuitive way.

3. Translate the following into Chinese

In the past decades, numerous schemes have been proposed to address these problems of conventional DTC. Many of them employ space vector modulation (SVM) to produce continuous voltage vectors, which can adjust the torque and flux more accurately and moderately, hence the torque and flux ripples were reduced while obtaining fixed switching frequency. Another merit of using SVM is that the sampling frequency doesnot need to be as high as that in conventional DTC. Various methods are proposed to obtain the commanding voltage vector, including deadbeat control, sliding mode control, PI controller etc. In the SVM-based DTC schemes, rotary coordinate transformation is often needed, which is more computationally intensive than the conventional DTC.

Part Ⅳ Power System and its Automation

Unit 1 Fundamentals of Power Systems

Text A Power of Single-phase Circuits

In the steady-state, most power system voltages and currents are (at least approximately) sinusoidal functions of time all with the same frequency. We are therefore very interested in sinusoidal steady-state analysis using phasors, impedances, admittances, and complex power. Some of these sinusoidal steady-state relations extend to an important class of transients as well.

We now investigate the instantaneous power absorbed by general RLC loads. We also introduce the concepts of real power, power factor, and reactive power.

1. Instantaneous Power

Power is the rate of change of energy with respect to time. The unit of power is a watt, which is a joule per second. Instead of saying that a load absorbs energy at a rate given by the power, it is common practice to say that a load absorbs power.

The instantaneous power in watts absorbed by an electrical load is the product of the instantaneous voltage across the load in volts and the instantaneous current into the load in amperes.

For a general load composed of RLC elements under sinusoidal-steady-state excitation, Assume that the load voltage is

$$u(t) = U_{\max}\cos(\omega t + \delta)\ \text{V} \tag{4-1}$$

And the load current is of the form

$$i(t) = I_{\max}\cos(\omega t + \beta)\ \text{A} \tag{4-2}$$

The instantaneous power absorbed by the load is then

$$\begin{aligned} p(t) &= u(t)i(t) = U_{\max}I_{\max}\cos(\omega t + \delta)\cos(\omega t + \beta) \\ p(t) &= UI\cos(\delta - \beta)\{1 + \cos[2(\omega t + \delta)]\} + UI\sin(\delta - \beta)\sin[2(\omega t + \delta)] \end{aligned} \tag{4-3}$$

Letting $I\cos(\delta - \beta) = I_R$ and $I\sin(\delta - \beta) = I_X$ gives

$$p(t) = \underbrace{UI_R\{1 + \cos[2(\omega t + \delta)]\}}_{p_R(t)} + \underbrace{UI_X\sin[2(\omega t + \delta)]}_{p_X(t)} \tag{4-4}$$

As indicated by Equation (4-4), the instantaneous power absorbed by the load has two components: One can be associated with the power $p_R(t)$ absorbed by the resistive component of the load, and the other can be associated with the power $p_X(t)$ absorbed by reactive (inductance or capacitive) component of the load. The phase angle $(\delta - \beta)$ represents the angle between the voltage and current.

2. Real Power

Equation (4-4) shows that the instantaneous power $p_R(t)$ absorbed by the resistive component of the load is a double-frequency sinusoid with average value P given by

$$P = UI_R = UI\cos(\delta - \beta) \text{ W} \tag{4-5}$$

The average power P is also called real power or active power. All three terms indicate the same quantity P given by Equation (4-5).

3. Power Factor

The term $\cos(\delta-\beta)$ in Equation (4-5) is called the power factor. The phase angle $(\delta-\beta)$, which is the angle between the voltage and current, is called the power factor angle. For DC circuits, the power absorbed by a load is the product of the DC load voltage and the DC load current; for AC circuits, the average power absorbed by a load is the product of the root mean square value (rms) load voltage U, rms load current I, and the power factor $\cos(\delta-\beta)$. For inductive loads, the current lags the voltage, which means β is less than δ, and the power factor $p_R(t)$ is said to be lagging. For capacitive loads, the current leads the voltage, which means β is greater than δ, and the power factor $\cos(\delta-\beta)$ is said to be leading. By convention, the power factor $\cos(\delta-\beta)$ is positive. If $|\delta-\beta|$ is greater than 90°, then the reference direction for current may be reversed, resulting in a positive value of $\cos(\delta-\beta)$.

4. Reactive Power

The instantaneous power absorbed by the reactive part of the load, given by the component $p_X(t)$ in Equation (4-4), is a double-frequency sinusoid with zero average value and with amplitude Q given by

$$Q = UI_X = UI\sin(\delta - \beta) \text{ var} \tag{4-6}$$

The term Q is given the name reactive power. Although it has the same units as real power, the usual practice is to define units of reactive power as volt-amperes reactive, or var.

5. Physical Significance of Real and Reactive Power

The physical significance of real power P is easily understood. The total energy absorbed by a load during a time interval T, consisting of one cycle of the sinusoidal voltage, is PT watt-seconds (Ws). During a time interval of n cycles, the energy absorbed is $P(n\mathrm{T})$ watt-seconds, all of which is absorbed by the resistive component of the load. A kilowatt-hour meter is designed to measure the energy absorbed by a load during a time interval (t_2-t_1), consisting of an integral number of cycles, by integrating the real power P over the time interval (t_2-t_1).

The physical significance of reactive power Q is not as easily understood. Q refers to the maximum value of the instantaneous power absorbed by the reactive component of the load. The instantaneous reactive power, given by the second term in $p_X(t)$ in Equation (4-4), is alternately positive and negative, and it expresses the reversible flow of energy to and from the reactive component of the load. Q may be positive or negative, depending on the sign of $(\delta-\beta)$ in Equation (4-6). Reactive power Q is a useful quantity when describing the operation of power systems.

Words and Phases

phasors *n.* 相量

admittances *n.* 导纳

instantaneous power 瞬时功率

joule *n.* 焦耳

product *n.* 乘积

reactive power 无功功率

active power 有功功率

root mean square value 均方根值

by convention 根据惯例

kilowatt-hour meter 电能表

integral *n.* 积分

shunt capacitors 并联电容器

Notes

The instantaneous reactive power, given by the second term in $p_X(t)$ in Equation (4-4), is alternately positive and negative, and it expresses the reversible flow of energy to and from the reactive component of the load.

式（4-4）中的第二部分 $p_X(t)$ 表示的瞬时无功功率是正负交替变化的，这表明电源和负载的无功部分之间的能量传递是可逆的。

Exercises

1. Answer the following questions according to the text

(1) How many components does the instantaneous power absorbed by the load consist of? What are they?

(2) What does "real power" mean?

(3) What is the relationship between the current and the voltage for inductive loads?

(4) How to definethe average power absorbed by a load for AC circuits?

(5) What is the physical significance of reactive power Q?

2. Translate the following sentences into Chinese according to the text

(1) Instead of saying that a load absorbs energy at a rate given by the power, it is common practice to say that a load absorbs power.

(2) The instantaneous power in watts absorbed by an electrical load is the product of the instantaneous voltage across the load in volts and the instantaneous current into the load in amperes.

(3) For AC circuits, the average power absorbed by a load is the product of the root mean square value (rms) load voltage U, rms load current I, and the power factor $\cos(\delta-\beta)$.

(4) For capacitive loads, the current leads the voltage, which means β is greater than δ, and the power factor $\cos(\delta-\beta)$ is said to be leading.

(5) A kilowatt-hour meter is designed to measure the energy absorbed by a load during a

time interval (t_2-t_1), consisting of an integral number of cycles, by integrating the real power P over the time interval (t_2-t_1).

3. Translate the following into Chinese

For a general load composed of RLC elements, the real power absorbed by a passive load is always positive. The reactive power absorbed by a load may be either positive or negative. When the load is inductive, the current lags the voltage, which means β is less than δ, and the reactive power absorbed is positive. When the load is capacitive , the current leads the voltage, which means β is greater than δ, and the reactive power absorbed is negative; or, alternatively, the capacitive load delivers positive reactive power.

Text B Advantages of Balanced Three-phase System versus Single-phase Systems

Fig. 4. 1 shows three separate single-phase systems. Each single-phase system consists of the following identical components: ① a generator represented by a voltage source and a generator impedance Z_g; ② a forward and return conductor represented by two series line impedances Z_L; ③ a load represented by an impedance Z_Y. The three single-phases systems, completely separated, are drawn in a Y configuration in the figure to illustrate two advantages of three-phases systems.

Each separate single-phase system requires that both the forward and return conductors have a current capacity (or ampacity) equal to or greater than the load current. However, if the source and load neutrals in Fig. 4. 1 are connected to form a three-phases system, and if the source voltages are balanced with equal magnitudes and with 120° displacement between phases, then the neutral current will be zero and the three neutral conductors can be removed. Thus, the balanced three-phase system, while delivering the same power to the three load impedances Z_Y, requires only half the number of conductors needed for the three separate single-phase systems. Also, the total I^2R line losses in the three-phase system are only half those of the three separate single-phase systems, and the line-voltage drop between the source and load in the three-phase system is half that of each single-phase system. Therefore, one advantage of balanced three-phase systems over separate single-phase systems is reduced capital and operating costs of transmission and distribution, as well as better voltage regulation.

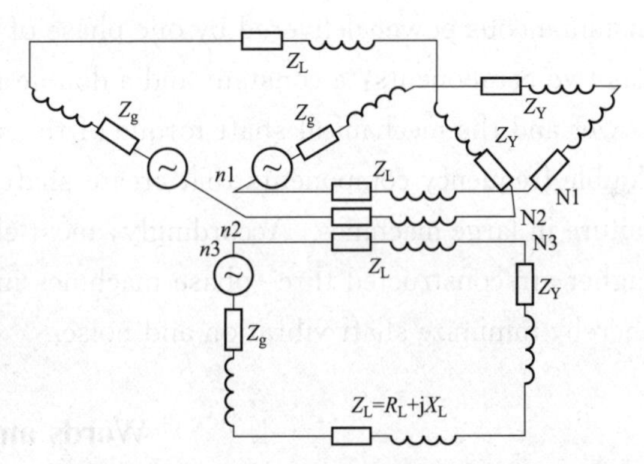

Fig. 4. 1 Three separate single-phase systems

Some three-phase systems such as Δ-connected systems and three-wire Y-connected systems do not have any neutral conductor. However, the majority of three-phase systems are four-wire Y-connected systems, where a grounded neutral conductor is used. Neutral conductors are used to transient overvoltages, which can be caused by lightning strikes and by line-switching operations, and to carry unbalanced currents, which can during unsymmetrical short-circuit conditions. Neutral conductors for transmission lines are typically smaller in size and ampacity than the phase conductors because the neutral current is nearly zero under normal operating conditions. Thus, the cost of a neutral conductor is substantially less than that of a phase conductor. The capital and operating costs of three-phase transmission and distribution systems with or without neutral conductors are substantially less than those of separate single-phase systems.

A second advantage of three-phase systems is that the total instantaneous electric power delivered by a three-phase generator under balanced steady-state conditions is (nearly) constant. A three-phase generator (constructed with its field winding on one shaft and with its three-phase windings equally displaced by 120° on the stator core) will also have a nearly constant mechanical input power under balanced steady-state conditions, since the mechanical input power equals the electrical output power plus the small generator losses. Furthermore, the mechanical shaft torque, which equals mechanical input power divided by mechanical radian frequency ($T_{mech}=P_{mech}/\omega_m$) is nearly constant.

On the other hand, the equation for the instantaneous electric power delivered by a single-phase generator under balanced steady-state conditions is the same as the instantaneous power delivered by one phase of a three-phase generator. Instantaneous power has two components: a constant and a double-frequency sinusoid. Both the mechanical input power and the mechanical shaft torque of the single-phase generator will have corresponding double-frequency components that create shaft vibration and noise, which could cause shaft failure in large machines. Accordingly, most electric generators and motors rated 5 kVA and higher are constructed three-phase machines in order to produce nearly constant torque and thereby minimize shaft vibration and noise.

Words and Phases

current capacity 载流容量
three-wire *n.* 三线制
four-wire *n.* 四线制
grounded neutral conductor 接地中性线
overvoltages *n.* 过电压
lightning strike *n.* 雷击
line-switching *n.* 线路转换
radian frequency 角频率
noise *n.* 噪声

Notes

1. Therefore, one advantage of balanced three-phase systems over separate single-phase

systems is reduced capital and operating costs of transmission and distribution, as well as better voltage regulation.

因此，三相平衡电路相对于单相电路的优势就如同好的电压调节一样，可以减少系统在传输和配电过程中的运行成本。

2. Neutral conductors are used to transient overvoltages, which can be caused by lightning strikes and by line-switching operations, and to carry unbalanced currents, which can during unsymmetrical short-circuit conditions.

中性线用来承受由于雷击、线路转换以及不对称短路运行所造成的不平衡电流等引起的瞬态过电压。

3. Accordingly, most electric generators and motors rated 5 kVA and higher are constructed three-phase machines in order to produce nearly constant torque and thereby minimize shaft vibration and noise.

因此，容量在5kVA及以上的发电机和电动机都设计成三相电机，其目的是为了产生一个恒定的转矩，由此可以减少电机轴的震动和噪声。

Exercises

1. Answer the following questions according to the text

(1) For the current capacity of each separate single-phase system, what relation is between the forward and return conductors and the load current?

(2) Ina three-phases system when the source voltages are balanced with equal magnitudes and with 120° displacement between phases, how much is the neutral current?

(3) Whose line-voltage drop is the bigger between the source and load in the three-phase system and each single-phase system?

(4) Why the most electric generators and motors rated 5 kVA and higher are constructed three-phase machines?

(5) Whichare the advantages of the balanced three-phase systems over separate single-phase systems?

2. Translate the following sentences into Chinese according to the text

(1) The three single-phases systems, completely separated, are drawn in a Y configuration in the figure to illustrate two advantages of three-phases systems.

(2) Each separate single-phase system requires that both the forward and return conductors have a current capacity (or ampacity) equal to or greater than the load current.

(3) However, if the source and load neutrals in Fig. 4. 1 are connected to form a three-phases system, and if the source voltages are balanced with equal magnitudes and with 120° displacement between phases, then the neutral current will be zero and the three neutral conductors can be removed.

(4) A second advantage of three-phase systems is that the total instantaneous electric power delivered by a three-phase generator under balanced steady-state conditions is (nearly)

constant.

(5) Both the mechanical input power and the mechanical shaft torque of the single-phase generator will have corresponding double-frequency components that create shaft vibration and noise, which could cause shaft failure in large machines.

3. Translate the following into Chinese

It is appropriate to mention that a balanced Δ-connected load is more common than a balanced Y-connected load. This is due to the ease with which loads may be added or removed from each phase of a Δ-connected load. This is very difficult with a Y-connected load because the neutral may not be accessible. On the other hand, Δ-connected sources are not common in practice because of the circulating current that will result in the Δ-mesh if the three-phase voltage are slightly unbalanced.

Unit 2 Power System Stability

Text A Power Flow Analysis

Power system stability refers to the ability of synchronous machines to move from one steady-state operating point following a disturbance to another steady-state operating point, without losing synchronism. There are three types of power system stability: steady-state, transient, and dynamic.

Steady-state stability involves slow or gradual changes in operating points. Steady-state stability studies, which are usually performed with a power flow computer program, ensure that phase angles across transmission lines are not too large, that bus voltages are close to nominal values, and that generators, transmission lines, transformers, and other equipment are not overloaded.

The model is appropriate for solving for the steady-state powers and voltages of the transmission system. The calculation is analogous to the familiar problem of solving for the steady-state voltages and currents in a circuit and is just as fundamental. It is an integral part of most studies in system planning and operation and is, in fact, the most common of power system computer calculations.

In power flow analysis the transmissions system is modeled by a set of buses or nodes interconnected by transmission links. Generators and loads, connected to various nodes of the system inject and remove power from the transmission system. To suggest the variety of possible studies, consider the system with the one-line diagram shown in Fig. 4. 2. The systems considered by power engineers would usually be larger, with as many as thousands of buses and thousands of transmission links.

In the Fig. 4. 2, the S_{Gi} are the injected (complex) generator powers and the S_{Di} are the (complex) load powers. The U_i are the complex (phasor) bus voltages. Transformers are assumed to have been absorbed into the generator, load, or transmission-line models and are not shown explicitly. It should be understood that we are restricting our attention to the transmission system, which transmits the power from the generators to the power substations. Thus the load powers shown in Fig. 4. 2 represent the power loads supplied to large industrial consumers and/or to a "subtransmission" system for further dispersal to distributions substations and ultimately a

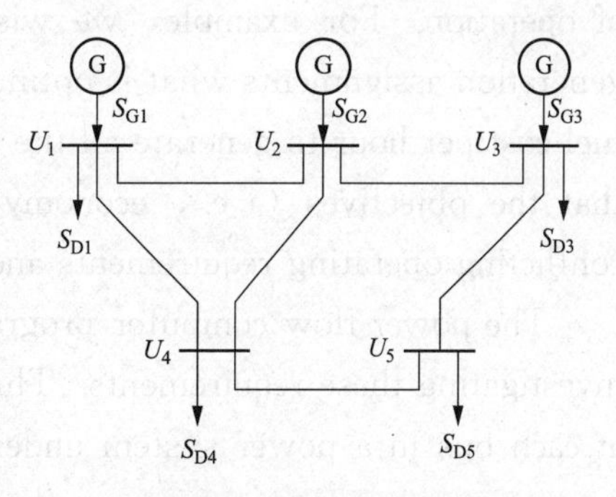

Fig. 4. 2 The system with one-line diagram

network of distribution feeders. While we are considering only the top layer of a multilayer system (the backbone of the overall system), it should be noted that the techniques developed in this section are applicable to the different layers of the system. The reason we concentrate on the transmission system is its basic importance and the fact that there are some interesting and vital problems, such as stability, unique to it.

The purpose of a power system is to deliver the power the customers in real time, on demand, within acceptable voltage and frequency limits, and in a reliable and economic manner. We are concerned here only with the operation and design of the system at the transmission.

In the analysis we assume that the load powers S_{Di}, are known constants. This assumption conforms to the driving nature of the customers, demand, wherein we may take it to be the input, and to the (usually) slowly varying nature of it, wherein we may take it to be constant. The effect of the actual variations, in S_{Di} with time can be studied by considering a number of different cases; for each we assume steady-state conditions. Frequently, the cases treated are the ones for which some difficulties in meeting system requirements may be expected. What are some of these difficulties? It may be the voltage magnitudes are not within acceptable limits or one or more lines are (thermally) overloaded, or that the stability margin for a transmission link is too small (i. e. , the power angle across a transmission link is too great), or that a particular generator is overloaded. Other studies relate to contingencies such as the emergency shutdown of a generator, or the loss of one or more transmission links due to equipment failure. With a given loading, the system may be functioning normally, but upon a single (or multiple) contingency outage the system is overloaded in some sense. In system operation it is desirable to operate the system in such a way that it is not overloaded in any way nor will it become so in the event of a likely emergency; in system planning there is a need to consider alternative plans to assure that these same objectives are met when the addition goes online.

In system operation and planning it is also extremely important to consider the economy of operation. For example, we wish to consider among all the possible allocations of generation assignments what is optimal in the sense of minimum production costs (i. e. , the fuel cost per hour to generate all the power needed to supply the loads). We note in passing that the objectives (i. e. , economy of operation and secure operation) frequently give conflicting operating requirements and compromises are usually required.

The power-flow computer program (commonly called load flow) is the basic tool for investigating these requirements. This program computers the voltage magnitude and angle at each bus in a power system under balanced three-phase steady-state conditions. It also computes real and reactive power flows for all equipment interconnecting the buses, as well as equipment losses. Both exciting power systems and proposed changes including new generation and transmission to meet projected load growth are of interest.

Conventional nodal or loop analysis is not suitable for power flow studies because the input data for loads are normally given in terms of power, not impedance. Also, generators are considered as power sources, not voltage or current sources. The power flow problem is therefore formulated as a set of nonlinear algebraic equations suitable for computer solution.

Words and Phases

stability *n.* 稳定性
disturbance *n.* 扰动，干扰
transient *n.* 暂态
dynamic *n.* 动态
gradual change 逐渐变化
power flow 潮流
bus *n.* 母线
nominal value 额定值
be analogous to 类似于
one-line diagram 单线图
substation *n.* 变电站
industrial consumer 工业用户
subtransmission *n.* 二次输电
distribution feeder 配电线路
conforms to 符合，遵照
contingency *n.* 临时情况，偶然
emergency shutdown 事故停机
outage *n.* 断电，停机
production costs 生产成本
nonlinear algebraic equation 非线性代数方程

Notes

1. Power system stability refers to the ability of synchronous machines to move from one steady-state operating point following a disturbance to another steady-state operating point, without losing synchronism.

功率系统的稳定性是指：当出现扰动时，同步电机在不失同步性的前提下，由一个稳定运行点过渡到另一个稳定运行点的能力。

2. Steady-state stability studies, which are usually performed with a powerflow computer program, ensure that phase angles across transmission lines are not too large, that bus voltages are close to nominal values, and that generators, transmission lines, transformers, and other equipment are not overload.

稳态分析通常指潮流计算，通过分析确保传输线的相角不会太大，母线电压接近于额定值以及发电机、传输线、变压器以及其他设备不会过载。

3. In power flow analysis the transmissions system is modeled by a set of buses or nodes interconnected by transmission links. Generators and loads, connected to various nodes of the system inject and remove power from the transmission system.

在进行潮流分析时，传输系统的模型用一组母线和在传输链上相互连接的节点来等效。连接到各个节点的发电机和负载分别向传输系统注入和吸收功率。

4. Thus the load powers shown in Fig. 4. 2 represent the power loads supplied to large industrial consumers and/or to a "subtransmission" system for further dispersal to distributions substations and ultimately a network of distribution feeders.

因此，图 4. 2 中的负载功率表示的是提供给那些大的工业用户或者是二次输电系统的负载，二次输电系统再把电能进一步分配给变电站和最终配电线路网。

Exercises

1. Answer the following questions according to the text

(1) What is the power system stability?

(2) How many types of power system stability are there? What are they?

(3) Which method is appropriate for solving for the steady-state powers and voltages of the transmission system?

(4) What is the purpose of a power system?

(5) In the power flow analysis, why is the load powers assumed a constants?

2. Translate the following sentences into Chinese according to the text

(1) Transformers are assumed to have been absorbed into the generator, load, or transmission-line models and are not shown explicitly.

(2) The reason we concentrate on the transmission system is its basic importance and the fact that there are some interesting and vital problems, such as stability, unique to it.

(3) The purpose of a power system is to deliver the power the customers in real time, on demand, within acceptable voltage and frequency limits, and in a reliable and economic manner.

(4) This assumption conforms to the driving nature of the customers, demand, wherein we may take it to be the input, and to the (usually) slowly varying nature of it, wherein we may take it to be constant.

(5) With a given loading, the system may be functioning normally, but upon a single (or multiple) contingency outage the system is overloaded in some sense.

3. Translate the following into Chinese

The power flow problem is the computation of voltage magnitude and phase angle at each bus in a power system under balanced three-phase steady-state conditions. As a by-product of this calculation, real and reactive power flows in equipment such as transmission lines and transformers, as well as equipment losses, can be computed.

The starting point for a power flow problem is a single-line diagram of the power system, from which the input data for computer solutions can be obtained. Input data consist of bus data, transmission line data, and transformer data.

Text B Transient Stability

Transient stability is the ability of the power system to maintain synchronism when subjected to severe transient disturbance. Stability depends on both the initial operating state of the system and the severity of the disturbance. Usually, the system is altered so that the post-disturbance steady-state operation differs from that prior to disturbance.

Transient stability, involves major disturbances such as loss of generation, line-switching operations, faults, and sudden load changes. Following a disturbance, synchronous machine frequencies undergo transient deviations from synchronous frequency and machine power angle change. The objective of transient stability study is to determine whether or not the machines will return to synchronous frequency with new steady-state power angles. Changes in power flow and bus voltage are also of concern.

The key question of transient stability is this : After a transient period, will the system lock back into a steady-state condition, maintaining synchronism? If it does, the system is said to be transient stable. If it does not, it is unstable and the system may break up into disconnected subsystems which in turn may experience further instability. Fig. 4. 3 shows an interesting mechanical analogy to the power system transient stability program. As shown in Fig. 4. 3, a number of masses representing synchronous machines are interconnected by a network of elastic strings representing transmission lines. Assume that this network is initially at rest in steady-state, with the net force on each string below its break point, when one of the strings is cut, representing the loss of a transmission line. As a result, the masses undergo transient oscillations and the forces on the string fluctuate. The system will then either settle down to a new steady-state operating point with a new set of string forces, or additional strings will break, resulting in an even weaker network and eventual system collapse. That is, for a given disturbance, the system is either transiently stable or unstable.

In today's large-scale power systems with many synchronous machines interconnected by complicated transmission networks, transient stability studies are best performed with a digital computer program. For a specified disturbance, the program alternately solves, step by step, algebraic power flow equations representing a network and nonlinear differential equations representing synchronous machines. Both predisturbance, disturbance, and post-disturbance computations are performed. The program output includes power angles and frequencies of synchronous machines, bus voltages, and power flows versus time.

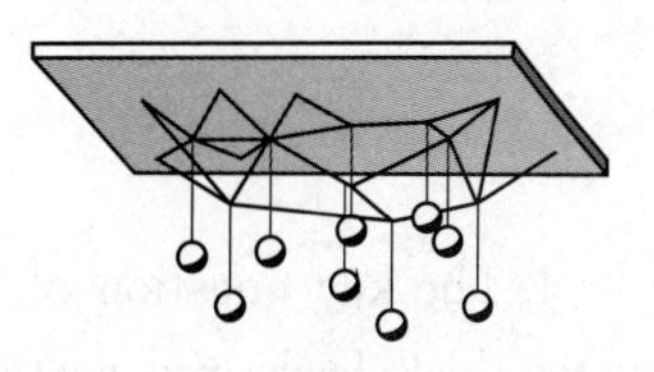

Fig. 4. 3 Mechanical analog of power system transient stability

In many cases, transient stability is determined during the first swing of machine power angles following a disturbance. During the first swing, which typically lasts about 1 s, the mechanical output power and the internal voltage of a generating unit are often assumed constant. However, where multiple swings lasting several seconds are of concern, models of turbine generator and excitation systems as well as more detailed machine models can be employed to obtain accurate transient stability results over the longer time period.

Dynamic stability involves an even longer time period, typically several minutes. It is possible for controls to affect dynamic stability even though transient stability is maintained. The action of turbine generator, excitation systems, tap-changing transformers, and controls from a power system dispatch center can interact to stabilize or destabilize a power system several minutes after a disturbance has occurred.

To simplify transient stability studies, the following assumptions are made:

(1) Only balanced three-phase systems and balanced disturbances are considered. Therefore, only positive-sequence networks are employed.

(2) Deviations of machine frequencies from synchronous frequency are small, and DC offset currents and harmonics are neglected. Therefore, the network of transmission lines, transformers, and impedance loads is essentially in steady-state; and voltages, currents, and powers can be computed from algebraic power flow equations.

Words and Phases

transient stability 暂态稳定
excursion *n.* 偏移
loss of generation 切除发电
fault *n.* 故障
transient deviation 瞬时偏差
mass *n.* 块，团
string *n.* 绳，带，线
break point 断点，断裂点
specified *n.* 特定的
differential equations 微分方程
first swing 第一摆动
dynamic stability 动态稳定性
turbine generator 汽轮发电机
offset *n.* 偏移量
tap-changing transformers 轴头转换变压器
dispatch center 调度中心

Notes

1. The key question of transient stability is this : After a transient period, will the system lock back into a steady-state condition, maintaining synchronism? If it does, the system is said to be transient stable. If it does not, it is unstable and the system may break up into disconnected subsystems which in turn may experience further instability.

暂态稳定的关键在于：系统在经过瞬变之后能否重新回到稳定状态并保持同步；如果能，那么系统可以说是一个暂态稳定系统。若不能，那么系统就是不稳定的，并且有可能分裂成互不相干的子系统，而这些子系统反过来会进一步影响系统的稳定性。

2. For a specified disturbance, the program alternately solves, step by step, algebraic power flow equations representing a network and nonlinear differential equations representing synchronous machines.

对于一个特定的扰动，计算程序可以采用多种方法，如阶梯法、代表电力网的潮流代数方程法以及代表同步电机的非线性微分方程法等。

Exercises

1. Answer the following questions according to the text

(1) What does transient stability depend on?

(2) What is transient stability of the power system?

(3) What is the objective of transient stability study?

(4) In today's large power systems, what are used study transient stability?

(5) To simplify transient stability studies, what assumptions are made?

2. Translate the following sentences into Chinese according to the text

(1) Stability depends on both the initial operating state of the system and the severity of the disturbance.

(2) Transient stability, involves major disturbances such as loss of generation, line-switching operations, faults, and sudden load changes.

(3) The system will then either settle down to a new steady-state operating point with a new set of string forces, or additional strings will break, resulting in an even weaker network andan eventual system collapse.

(4) Following a disturbance, synchronous machine frequencies undergo transient deviations from synchronous frequency and machine power angle change.

(5) The action of turbine generator, excitation systems, tap-changing transformers, and controls from a power system dispatch center can interact to stabilize or destabilize a power system several minutes after a disturbance has occurred.

3. Translate the following sentences into Chinese

A lightning stroke hits a transmission line, "breaking down" the air between a pair of conductors, or from line to ground and creating an ionized path. Sixty-Hertz line currents also flow trough this ionized path, maintaining the ionization even after the lightning stroke energy has been dissipated. This constitutes a short circuit or "fault" as real as if the conductors were in physical contact. To remove the fault, relays detect the short circuit and cause circuit breakers to open at both ends of the line. The line is deenergized and the air deionizes, reestablishing its insulating properties. We say that the fault "clears". The circuit breakers are set to reclose automatically after a preset interval of time, reestablishing the original circuit. This sequence of events (called a fault sequence) constitutes a shock to the power system and is accompanied by a transient.

Unit 3 Electrical Power System

Text A Components of Power System

Fig. 4. 4 shows a minimum power system. The system consists of an energy source, a prime mover, a generator, and a load. The energy source may be coal, gas, or oil burned in a furnace to heat water and generator steam in a boiler. The prime mover may be a steam turbine, a hydraulic turbine or water wheel, or an internal combustion engine. Each one of these prime movers has the ability to convert energy in the form of heat, falling water, or fuel into rotation of a shaft, which in turn will drive the generator. The electrical load on the generator may be lights, motors, heaters, or other devices, alone or in combination. The control system functions to keep speed of the machines substantially constant and the voltage within prescribed limits, even the load may change. The control system may include a main stationed in the power plant who watches a set of meters on the generator-output terminals and makes the necessary adjustments manually. In a modern station, the control system is a servomechanism that senses a generator-output conditions and automatically makes the necessary changes in energy input and field current to hold the electrical output within certain specifications.

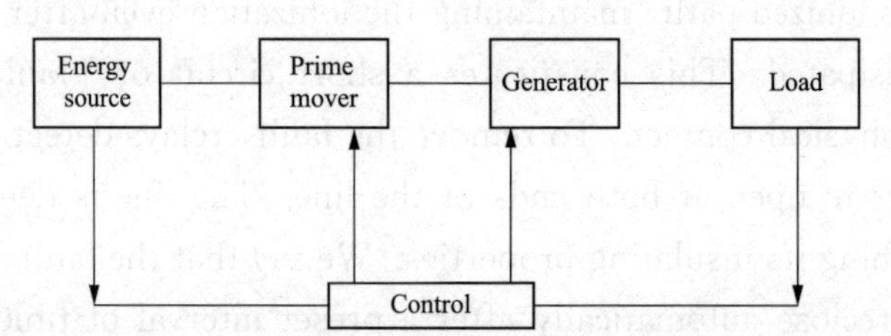

Fig. 4. 4 Minimum power system

Modern power systems are usually large-scale, geographically distributed, and with hundreds to thousands of generators operating in parallel and synchronously. They may vary in size and structure from one to another, but they all have basic characteristics:

(1) Are comprised of three-phase AC systems operating essentially at constant voltage. Generation and transmission facilities use three-phase equipment. Industrial loads are invariably three-phase; single-phase residential and commercial loads are distributed equally among the phases so as to effectively form a balanced three-phase system.

(2) Use synchronous machines for generation of electricity. Prime movers convert the primary energy (fossil, nuclear, and hydraulic) to mechanical energy that is, in turn, converted to electrical energy by synchronous generators.

(3) Transmit power over significant distances to consumers spread over a wide area. This requires a transmission system comprising subsystems operating at different voltage levels.

Fig. 4. 5 shows a basic element of a modern power system. Electric power is produced at generating stations (GS) and transmitted to consumers through a complex network of individual components, including transmission lines, transformers, and switching devices. It is common practice to classify the transmission network into the following subsystems: transmission system; subtransmission system; distribution system.

The transmission system interconnects all major generating stations and main load centers in the system. It forms the backbone of the integrated power system and operates at the highest voltage levels. The generator voltages are usually in the range of 11 kV to 35 kV. These are stepped up to the transmission voltage level, and power is transmitted to transmission substations where the voltages are stepped down to the subtransmission level (typically, 69 kV to 138 kV). The generation and transmission subsystems are often referred to as the bulk power system.

The subtransmission system transmits power in small quantities from the transmission substations to the distribution substations. Large industrial customers are commonly supplied directly from the subtransmission system. In some systems, there is no clear demarcation between subtransmission and transmission circuits. As the system expands and higher voltage levels become necessary for transmission, the older transmission lines are often relegated to subtransmission function.

The distribution system represents the final stage in the transfer of power to the individual customers. The primary distribution voltage is typically between 4. 0 kV and 34. 5 kV. Small industrial customers aresupplied by primary feeders at this voltage level. The secondary distribution feeds supply residential and commercial customers at 120/240 V.

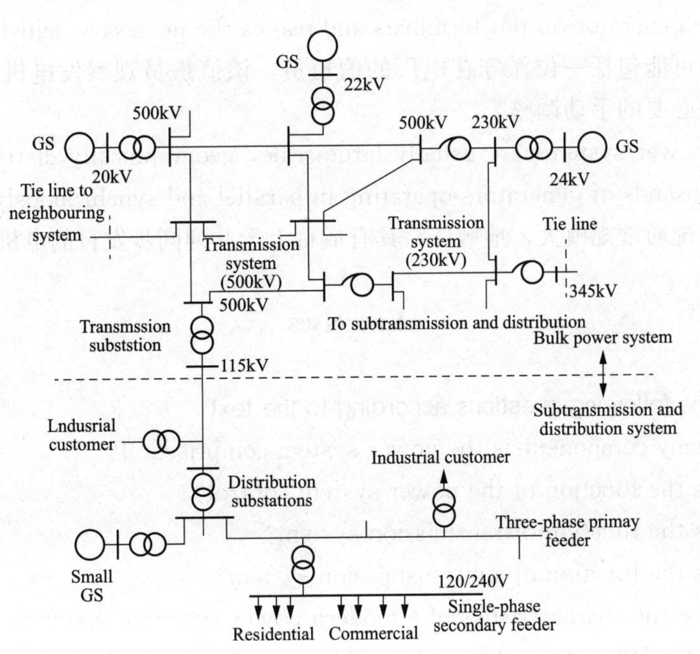

Fig. 4. 5　Basic elements of a modern power system

Small generating plants located near the load are also connected the subtransmission or distribution system directly. Interconnections to neighboring power systems are usually formed at the transmission system level. The overall system thus consists of multiple generating sources and several layers of transmission networks. This provides a high degree of structural redundancy that enables the system to withstand unusual contingencies without service disruption to the customers.

Words and Phases

fossil *n.* 化石，石块

prime mover 原动机

steam turbine 汽轮机

hydraulic turbine 水轮机

internal combustion engine 内燃机

servomechanism *n.* 伺服机构

mechanical energy 机械能

bulk power system 大容量电力系统

contingencies *n.* 偶然事故，意外事故

Notes

1. The control system functions to keep speed of the machines substantially constant and the voltage within prescribed limits, even the load may change.

控制系统的作用是在负载变化的情况下仍能保持电机的基本稳定并将电压控制在规定的范围之内。

2. The control system may include a man stationed in the power plant who watches a set of meters on the generator-output terminals and makes the necessary adjustments manually.

控制系统也可能包括一位派守在电厂的值班员，该值班员观察发电机输出端的一套仪表，并做出一些必要的手动调整。

3. Modern power systems are usually large-scale, geographically distributed, and with hundreds to thousands of generators operating in parallel and synchronously.

现代电力系统通常规模大，地域广，具有成百上千并列同步发行的电机组。

Exercises

1. Answer the following questions according to the text

(1) How many componentsis the power system comprised of?

(2) What is the function of the power system control?

(3) What is the function oftransmission system?

(4) What is the function of subtransmission system?

(5) What are the characteristics of a modern power system in network structure?

2. Translate the following sentences into Chinese according to the text

(1) Each one of these prime movers has the ability to convert energy in the form of

heat, falling water, or fuel into rotation of a shaft, which in turn will drive the generator.

(2) In a modern station, the control system is a servomechanism that senses a generator-output condition and automatically makes the necessary changes in energy input and field current to hold the electrical output within certain specifications.

(3) Industrial loads are invariably three-phase, single-phase residential and commercial loads are distributed equally among the phases so as to effectively form a balanced three-phase system.

(4) Electric power is produced at generating stations (GS) and transmitted to consumers through a complex network of individual components, including transmission lines, transformers, and switching devices.

(5) This provides a high degree of structural redundancy that enables the system to withstand unusual contingencies without service disruption to the customers.

3. Translate the following sentences into Chinese

In planning an electric utility system, the physical location of the generating station, transmission lines and substations must be carefully planned to arrive at an acceptable, economic solution. We can sometimes locate a generating station next to the source of energy (such as a coal mine) and use transmission lines to carry the electrical energy to where it is needed. When this is neither practical economical, we have to transport the energy (coal, gas, oil) by ship, train, or pipeline to the generating station. The generating station may, therefore, be near to, or far from, the ultimate user of the electrical en energy. Some of the obstacles prevent transmission lines from following the shortest route. Due to these obstacles, both physical and legal, transmission lines often follow a zigzag path between the generating station and the ultimate user.

Text B Electrical Energy of Power System

Electricity is only one of many forms of used in industry, homes, businesses, and transportation. It has many desirable features; it is clean (particularly at the point of use), convenient, relatively easy to transfer from point of source to point of use, and highly flexible in its use. In some cases it is an irreplaceable source of energy.

1. Demand of an Electrical System

The totalpower drawn by the customers of a large utility system fluctuates between wide limits, depending on the seasons and time of day. Fig. 4. 6 shows how the system demand (power) varies during a typical day in the summer and a typical day in the winter. The pattern of the daily demand is remarkably similar for the two seasons. During the winter the peak demand of 15 GW (= 15000 MW) is higher than the summer peak of 10 GW. Nevertheless, both peaks occur about 17 : 00 because increased domestic activity at this time coincides with industrial and commercial centers that are still operating at full capacity.

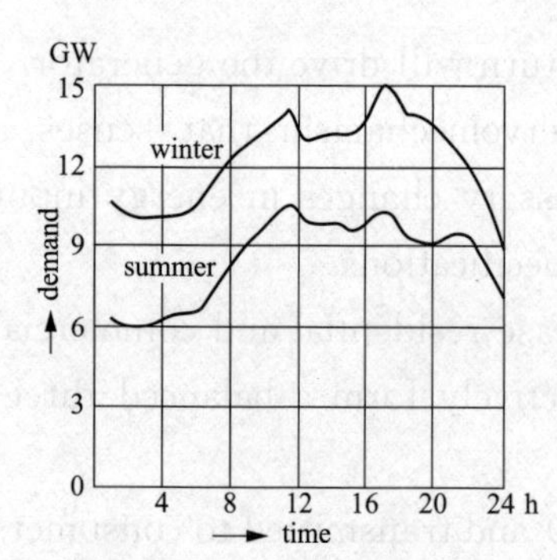

Fig. 4. 6 Demand curve of a large system during a summer day and a winter day

The load curve of Fig. 4. 6 shows the seasonal variations for the same system. Note that the peak demand during the winter (15 GW) is more than twice the minimum demand during the summer (6 GW).

In examining the curve, we note that the demand throughout the year never falls below 6 GW. This is the base load of the system. We also see that the annual peak load is 15 GW. The base load has to be fed 100 percent of the time, but the peak load may occur for only 0.1 percent of the time. Between these two extremes, we have intermediate loads that have to be fed for less than 100 percent of the time.

If we plot the duration of each demand on an annual base, we obtain the load duration curve of Fig. 4. 7. For example, the curve shows that a demand of 9 GW lasts 70 percent of the time, while a demand of 12 GW lasts for only 15 percent of the time. The graph is divided into base, intermediate, and peak load sections. The peak-load portion usually includes demands that last for less than 15 percent of the time. On this basis the system has to deliver 6 GW of base power, another 6 GW of intermediate power, and 3 GW of peak power.

These power blocks give rise to three types of generating stations:

(1) Base-power stations that deliver full power all times: nuclear stations and coal-fired stations are particularly well adapted to furnish base demand.

(2) Intermediate-power stations that can respond relatively quickly to changes in demand, usually by adding or removing one or more generating units: hydropower stations are well adapted for this purpose.

(3) Peak-generating stations that deliver power for brief intervals during the day: such stations must be put into service very quickly. Consequently, they are equipped with prime movers such as diesel engines, gas turbines, compressed-air motors, or pumped-storage turbines that can be started up in a few minutes. In this regard, it is worth mentioning that thermal generating stations using gas or coal take from 4 to 8 hours to start up, while nuclear stations may take several days. Obviously, such generating stations cannot be used to supply short term peak power.

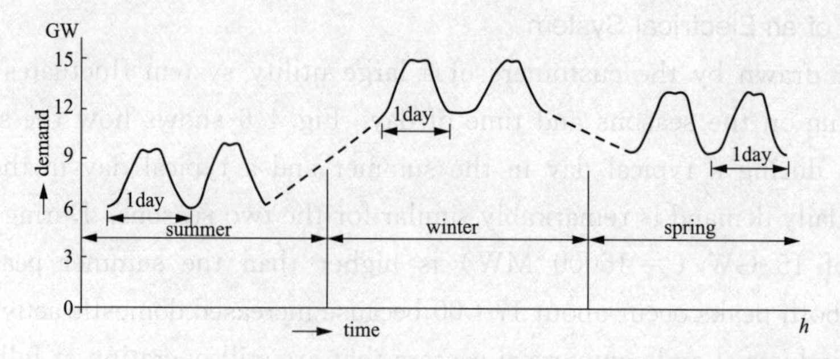

Fig. 4. 7 Demand curve of a large electric utility system during one year

2. Location of the Generating Station

In planning an electric utility system, the physical location of the generating station, transmission lines and substations must be carefully planned to arrive at an acceptable, economic solution. We can sometimes locate a generating station next to the source of energy (such as a coal mine) and use transmission lines to carry the electrical energy to where it is needed. When this is neither practical economical, we have to transport the energy (coal, gas, oil) by ship, train, or pipeline to the generating station. The generating station may, therefore, be near to, or far from, the ultimate user of the electrical en energy. Some of the obstacles prevent transmission lines from following the shortest route. Due to these obstacles, both physical and legal, transmission lines often follow a zigzag path between the generating station and the ultimate user.

3. Types of Generating Station

There are three main types of generating stations:

(1) Thermal generating stations

(2) Hydro power generating station

(3) Nuclear generating stations

Although we can harness the wind, tide, and solar energy, these energy sources represent a tiny part of the total energy we need.

Words and Phases

irreplaceable *adj*. 不能替代的	gas turbine 燃气轮机
fluctuate *v*. 波动，变化	compressed-air motor 空气压缩机
nuclear station 核电站	pumped-storage turbine 抽水蓄能轮机
coal-firedstation 燃煤电站，热电站	zigzag *adj*."z"字型的，曲折的
hydropower station 水电站	thermal generating station 热电站，热电厂
diesel engine 柴油机	

Notes

The totalpower drawn by the customers of a large utility system fluctuates between wide limits, depending on the seasons and time of day.

一个大的用电系统的消费者的用电量具有很大的波动性，这主要是由于季节的变化以及一天内的不同时间段而造成的。

Exercises

1. Answer the following questions according to the text

(1) What are the desirable features of electricity?

(2) What time is the peakdemand during a typical day?

(3) Which stations are adapted for intermediate load?

(4) Which stations are adapted for peak load?

(5) What other types of energy source are there besides thermal, hydro, nuclear energy sources?

2. Translate the following sentences into Chinese according to the text

(1) It is clean (particularly at the point of use), convenient, relatively easy to transfer from point of source to point of use, and highly flexible in its use.

(2) Nevertheless, both peaks occur about 17∶00 because increased domestic activity at this time coincides with industrial and commercial centers that are still operating at full capacity.

(3) In this regard, it is worth mentioning that thermal generating stations using gas or coal take from 4 to 8 hours to start up, while nuclear stations may take several days.

(4) Due to these obstacles, both physical and legal, transmission lines often follow a zigzag path between the generating station and the ultimate user.

(5) Although we can harness the wind, tide, and solar energy, these energy sources represent a tiny part of the total energy we need.

3. Translate the following sentences into Chinese

The hydraulic resources of most modern countries are already fully developed. Consequently, we have to rely on the thermal and nuclear stations to supply the growing need for electrical energy.

Thermal generating stations produce electricity from the heat released by thecombustion of coal, oil, or natural gas. Most stations have ratings between 200 MW and 1500 MW so as to attain the high efficiency and economy of a large installation. Such a station has to be seen to appreciate its enormous complexity and size.

Thermal stations are usually located near a river or lake because large quantities of cooling water are needed to condense the steam as it exhausts from the turbines. The efficiency of thermal stations is always low because of the inherent low efficiency of the turbines.

Unit 4 Transmission of Electrical Energy

Text A Power Distribution System

The transmission of electrical energy does not usually raise as much interest as does its generation and utilization; consequently, we sometimes tend to neglect this important subject. This is unfortunate because the human and material resources involved in transmission are much greater than those employed in generation.

Electrical energy is carried by conductors such as overhead transmission lines and underground cable. Although these conductors appear very ordinary, they possess important electrical properties that greatly affect the transmission of electrical energy.

1. Principal Components of a Power Distribution System

In order to provide electrical energy to consumers in usable form, a transmission and distributions system must satisfy some basic requirements. Thus the system must:

- Provide, at all times, the power that consumers need.
- Maintain a stable, nominal voltage that does not vary by more than ±10%.
- Maintain a stable frequency that does not vary by more than ±0. 1 Hz.
- Supply energy at an acceptable price.
- Meet standards of safety.
- Respect environmental standards.

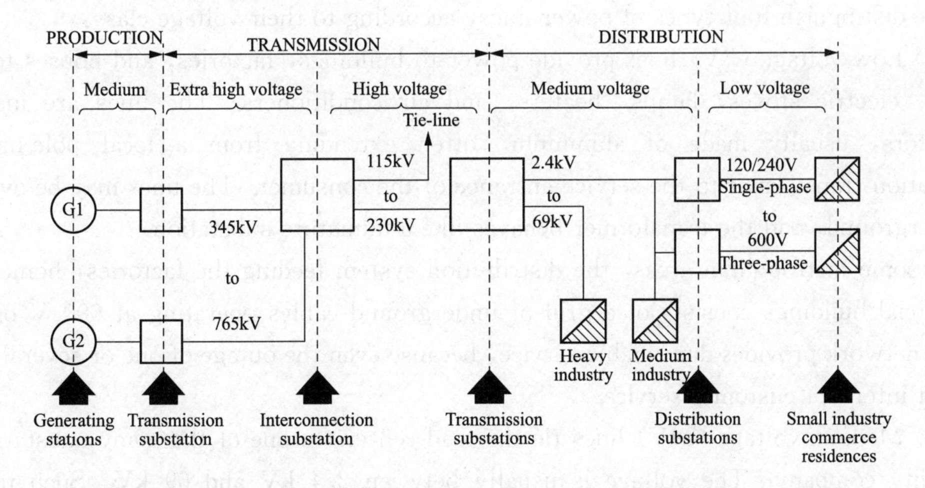

Fig. 4. 8 Single line diagram of a production, transmission, and distribution system

Fig. 4. 8 shows an elementary diagram of a transmission and distribution system. It consists of two generating stations G1 and G2, a few substations, an interconnecting substation and several commercial, residential, and industrial loads. The energy is carried over lines designated extra high voltage (EHV), high voltage (HV), medium voltage

(MV), and low voltage (LV).

Transmission substations (Fig. 4. 8) change the line voltage by means of step-up and step-down transformers and regulate it by means of static var compensators, synchronous condensers, or transformers with variable taps.

Distribution substations change the medium voltage to low voltage by means of step-down transformers, which may have automatic tap-changing capabilities to regulate the low voltage. The low voltage ranges from 120/240 V single phase to 600 V, three-phase. It serves to power private residences, commercial and institutional establishment and small industry.

Interconnecting substations tie different power systems together, to enable power exchanges between them, and to increase the stability of the overall network.

These substations also contain circuit breakers, fuses, and lightning arresters, to protect expensive apparatus, and to provide for quick isolation of faulted lines from the system. In addition, control apparatus, power measuring devices, disconnect switches, capacitors, inductors, and other devices may be part of a substation.

2. Types of Power Lines

The design of a power line depends upon the following criteria:

➢ The amount of active power it has to transmit.

➢ The distance over which the power must be carried.

➢ The cost of the power line.

➢ Esthetic considerations, urban congestion, ease of installation, and expected load growth.

We distinguish four types of power lines, according to their voltage class:

(1) Low voltage (LV) lines provide power to buildings, factories, and houses to drive motors, electric stoves, lamps, heaters, and air conditioners. The lines are insulated conductors, usually made of aluminum, often extending from a local pole-mounted distribution transformer to the service entrance of the consumer. The lines may be overhead or underground, and the transformer behaves like a miniature substation.

In some metropolitan areas, the distribution system feeding the factories, homes, and commercial buildings consists of a grid of underground cables operating at 600 V or less. Such a network provides dependable service, because even the outage of one or several cables will not interrupt customer service.

(2) Medium voltage (MV) lines tie the load centers to one of the many substations of the utility company. The voltage is usually between 2. 4 kV and 69 kV. Such medium voltage radial distribution systems are preferred in the larger cities. In radial systems the transmission lines spread out like fingers from one or more substations to feed power to various load centers, such as high-rise buildings, shopping centers, and campuses.

(3) High voltage (HV) lines connect the main substations to the generating stations. The lines are composed of aerial conductors or underground cables operating at voltages

below 230 kV. In this category we also find lines that transmit energy between two power systems, to increase the stability of the network.

(4) Extra high voltage (EHV) lines are used when generating stations are very far from the load centers. We put these lines in a separate class because of their special electrical properties. Such lines operate at voltages up to 800 kV and may be as long as 1000 km.

Words and Phases

material resources 物力
overhead *adj*. 架空的
transmission line 传输线
extra high voltage 超高压
Transmission substation 输电变电站
step-up transformer 升压变压器
step-down transformer 降压变压器
var compensator 无功补偿器
synchronous condenser 同步调相机
tap *n*. 抽头，分接头
fuse *n*. 熔断器
lightning arresters 避雷针
criteria *n*. 规范，标准
urban congestion 市区拥挤，城市过密化
aluminum *n*. 铝
miniature *n*. 小型的
metropolitan *adj*. 大城市的

Notes

1. The transmission of electrical energy does not usually raise as much interest as does its generation and utilization; consequently, we sometimes tend to neglect this important subject.

电能的传输并没有像它的产生和利用那样引起人们的重视，甚至有时我们会忽视它的重要性。

2. Transmission substations change the line voltage by means of step-up and step-down transformers and regulate it by means of static var compensators, synchronous condensers, or transformers with variable taps.

输电变电站通过升压和降压变压器来改变线路电压，通过无功补偿器、同步调相机以及变压器来调节电压。

Exercises

1. Answer the following questions according to the text

(1) What is usuallyneglected in the transmission of electrical energy?

(2) In order to provide electrical energy to consumers in usable form, what must be satisfied for a transmission and distributions system?

(3) Which typesof power lines are designated?

(4) What are used tochange the line voltage in transmission substations?

(5) What components are also contained for transmission and distribution system besides the

generating stations, a few substations, the interconnecting substation and loads?

2. Translate the following sentences into Chinese according to the text

(1) This is unfortunate because the human and material resources involved in transmission are much greater than those employed in generation.

(2) Electrical energy is carried by conductors such as overhead transmission lines and underground cable.

(3) Distribution substations change the medium voltage to low voltage by means of step-down transformers, which may have automatic tap-changing capabilities to regulate the low voltage.

(4) Interconnecting substations tie different power systems together, to enable power exchanges between them, and to increase the stability of the overall network.

(5) The lines are insulated conductors, usually made of aluminum, often extending from a local pole-mounted distribution transformer to the service entrance of the consumer.

3. Translate the following into Chinese

The fundamental purpose of a transmission or distribution lines is to carry active power from one point to another. If it also has to carry reactive power, the latter should be kept as small as possible. In addition, a transmission line should possess the following basic characteristic:

(1) The voltage should remain as constant as possible over the entire length of the line, from source to load, and for all loads between zero and rated load.

(2) The line losses must be small so as to attain a high transmission efficiency.

(3) The I^2R losses must not overheat the conductors.

If the line alone cannot satisfy the above requirements, supplement equipment, such as capacitors and inductors, must be added until the requirements are not.

Text B Direct-Current Transmission

1. Features of DC Transmission

The development of high-power, high-voltage electronic converters has made it possible to transmit and control large blocks of power using direct current. Direct-current transmission offers unique features that complement the characteristics of existing AC networks.

What are the advantages of transmitting power by DC rather than by AC? They may be listed as follows:

(1) DC power can be controlled much more quickly. For example, power in the megawatt range can be reversed in a DC line in less than one second. This feature makes it useful to operate DC transmission lines in parallel with existing AC networks. When instability is about to occur (due to a disturbance on the AC system), the DC power can be changed in amplitude to counteract and dampen out the power oscillations. Quick power control also means that circuit currents can be limited to much lower values than those encountered on AC networks.

(2) DC power can be transmitted in cables over great distances. We have seen that the capacitance of a cable limits AC power transmission to a few tens of kilometers. Beyond this limit, the reactive power generated by cable capacitance exceeds the rating of the cable itself. Because capacitance does not come into play under steady-stated conditions, there is theoretically no limit to the distance that power may be carried this way. As a result, power can be transmitted by cable under large bodies of water, where the use of AC cable is unthinkable. Furthermore, underground DC cable may be used to deliver power into large urban centers. Unlike overhead lines, underground cable is invisible, free from atmospheric pollution, and solves the problem of securing rights of way.

(3) We have seen that AC power can only be transmitted between centers operating at the same frequency. Furthermore, the power transmitted depends upon line reactance and the phase angle between the voltages at each end of the line. But when power is transmitted by DC, frequencies and phase angles do not come into the picture, and line reactance does not limit the steady-state power flow. If anything, it is only the resistance of the line that limits the flow. This also means that power can be transmitted over greater distances by using DC. However, this is a marginal benefit because large blocks of AC power are already being carried over distances exceeding 1000 km.

(4) Overhead DC transmission lines become economically competitive with AC lines when the length of the line exceeds several hundred kilometers. The width of the power corridor is less, and experience to date has shown that outages due to lightning are somewhat reduced. Consequently DC transmission lines are being used to carry bulk power directly from a generating station located near a coal mine or waterfall, to the load center.

(5) At the opposite extreme of great distance are back-to-back converters, which interconnect large adjacent AC systems with a DC transmission line that is only a few meters long. Back to back converters enable the two systems to operate at their respective frequencies and phase angles. As a result, disturbances on one system do not tend to destabilize the other system. Furthermore, the power flow between the systems can be modified and even reversed in a matter of milliseconds, which is faster than faster than on an AC system.

Unlike AC transmission lines, it is not easy to tap power off at different points along a DC line. In effect, DC lines are usually point-to-point systems, tying one large generating station to one large power-consuming center. Electronic converters are installed at each end of the transmission line, but none in between.

2. Basic DC Transmission System

ADC transmission system consists basically of a DC transmission line connecting two AC systems. A converter at one end of the line converts AC power into DC power while a similar converter at the other end reconverts the DC power into AC power. One converter acts therefore as a rectifier, the other as an inverter.

Stripped of everything but the bare essentials, the transmission system may be represented by the circuit of Fig. 4. 9. converter 1 is a three-phase, six-pulse rectifier that converts the AC power of line 1 into DC power. The DC power is carried over a two-conductor transmission line and reconverted to AC power by means of converter 2, acting as an inverter. Both the rectifier and inverter are line-commutated by the respective line voltages to which they are connected. Consequently, the networks can function at entirely different frequencies without affecting the power transmission between them.

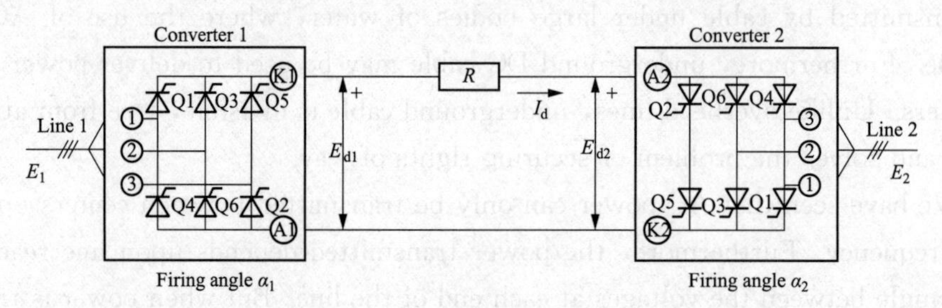

Fig. 4. 9 Elementary DC transmission system connecting three-phase line 1 to three-phase line 2

Power flow may be reversed by changing the firing angles α_1 and α_2, so that converter 1 becomes an inverter and converter 2 a rectifier. Changing the angles reverses the polarity of the conductors, but the direction of current flow remains the same. This mode of operation is required because thyristors can only conduct current in one direction.

The DC voltages E_{d1} and E_{d2} at each converter station are identical, except for the *IR* drop in the line. The drop is usually so small that we can neglect it, except insofar as it affects losses, efficiency, and conductor heating.

Due to the high voltages encountered in transmission lines, each thyristor shown in Fig. 4. 9 is actually composed of several thyristors connected in series. Such a group of thyristors is often called a valve. Thus, a valve for a 50 kV, 1000 A converter would typically be composed of 50 thyristors connected in series. Each converter in Fig. 4. 9 would, therefore, contain 300 thyristors. The 50 thyristors in each bridge arm are triggered simultaneously, so together they act like a super-thyristor.

Words and Phases

dampen out 减弱掉，吸收尽

atmospheric *adj*. 大气的

back-to-back *adj*. 紧贴的，背靠背的

adjacent *adj*. 相邻的，邻近的

destabilize *v*. 使动摇，使不稳定

millisecond *n*. 毫秒

insofar *adv*. 在……范围之内

Notes

1. Because capacitance does not come into play under steady-stated conditions, there is theoretically no limit to the distance that power may be carried this way.

在稳定条件下电容并不能起作用，因此，电缆到底能够输送多远的距离并没有一个理论

上的界限。

2. At the opposite extreme of great distance are back-to-back converters, which interconnect large adjacent AC systems with a DC transmission line that is only a few meters long.

在长距离输电的两端是相邻的变换器，该变换器用仅仅几米长的传输线连接两个邻近的交流系统。

Exercises

1. Answer the following questions according to the text

(1) How many advantages of transmitting power by DC are there than by AC? What are they?

(2) Why the capacitance of a cable limits AC power transmission to a few tens of kilometers?

(3) Which is faster on the power flow modify of between the DC transmission system and the AC transmission system?

(4) WhyDC lines are usually point-to-point systems?

(5) Why power flow may be reversed during changing the angles but the direction of current flow cannot be reversed?

2. Translate the following sentences into Chinese according to the text

(1) Quick power control also means that circuit currents can be limited to much lower values than those encountered on AC networks.

(2) Beyond this limit, the reactive power generated by cable capacitance exceeds the rating of the cable itself.

(3) But when power is transmitted by DC, frequencies and phase angles do not come into the picture, and line reactance does not limit the steady-state power flow.

(4) The width of the power corridor is less, and experience to date has shown that outages due to lightning are somewhat reduced.

(5) Consequently, the networks can function at entirely different frequencies without affecting the power transmission between them.

3. Translate the following into Chinese

Insulated cables have many applications in the field of electric power. Small cables are used as extension cords around offices, homes, and factories. Larger cables are used for connections tomachines that are movable over restricted distances. In some instances portable cables carry quite heavy electrical loads, as for example the case of electrically driven drag lines, which require several thousand horsepower for operation. Overhead cables find application in distribution circuits where tree conditions or proximity to buildings and other structures make the use of open-wire lines impracticable. Underground cables are used in many situations, including major transmission circuits between large stations. Some cable power circuits operate at 765 kV and carry loads of several hundred megawatts. Cables of even higher voltage will probably become available soon.

Unit 5 Faults and Protection on Power System

Text A Power System Faults

1. Faults and Its Damage

Faults are the unintentional or intentional connecting together of two or more conductors which ordinarily operate with a difference of potential between them. The connection between conductors may be physical metallic contact or it may be through an arc.

On the occurrence of a fault, current and voltage conditions become abnormal, the delivery of power from the generating stations to the loads may be unsatisfactory over a considerable area, and if the faulted equipment is not promptly disconnected from the remainder of the system, damage may result to other pieces of operating equipment. For example, at the fault point itself, there may be arcing, accompanied by high temperatures and, possibly, fire and explosion. There may be destructive mechanical forces due to very high currents. Overvoltages may stress insulation beyond the breakdown value. Even in the case of less severe faults, high currents in the faulted system may overheat equipment; sustained overheating may reduce the useful life of the equipment. Clearly, faults must be removed from the system as rapidly as possible. It is apparent that the power system designer must anticipate points at which faults may occur, be able to calculate conditions that exist during a fault, and provide equipment properly adjusted to open the switches necessary to disconnect the faulted equipment from the remainder of the system. Ordinarily it is desirable that no other switches on the system are opened, as such behavior would result in unnecessary modification of the system circuits.

2. Various Faults

Faults of many types and causes may appear on electric power systems. Many of us in our homes have seen frayed lamp cords which permitted the two conductors of the cord to come in contact with each other. When this occurs, there is a resulting flash, and if breaker or fuse equipment functions properly, the circuit is opened.

Overhead lines, for the most part, are constructed of bare conductors. These are sometimes accidentally brought together by action of wind, sleet, trees, cranes, airplanes, or damage to supporting structures. Overvoltages due to lighting or switching may cause flashover of supporting or from conductor to conductor. Contamination on insulators sometimes results in flashover even during normal voltage conditions.

The conductors of underground cables are separated from each other and from ground by solid insulation, which may be oil-impregnated paper or a plastic such as polyethylene. These materials undergo some deterioration with age, particularly if overloads on the cables have resulted in their operation at elevated temperature. Any small void present in the body of the insulating material will result in ionization of the gas contained therein, the products of which react unfavorably with the insulation. Deterioration of the insulation may result in failure of the material to retain its insulating properties, and short circuits will develop between the cable conductors. The possibility of cable failure is increased if lighting or switching produces transient voltage of abnormally high values between the conductors.

Transformer failures may be result of insulation deterioration combined with overvoltages due to lighting or switching transients. Short circuits due to insulation failure between adjacent turns of the same winding may result from suddenly applied overvoltages. Major insulation may fail, permitting arcs to be established between primary and secondary windings or between the winding and grounded metal parts such as the core or tank.

Generators may fail due to breakdown of the insulation between adjacent turns in the same slot, resulting in a short circuit in a single turn of the generator. Insulation breakdown may also occur between one of the windings and the grounded steel structure in which the coils are embedded. Breakdown between different windings lying in the same slot results in short-circuiting extensive sections of machine.

Balanced three-phase faults, like balanced three-phase loads, may be handled on a line toneutral basis or on an equivalent single-phase basis. Problems may be solved either in terms of volts, amperes, and ohms. The handing of faults on single-phase lines is of course identical to the method of handing three-phase faults on an equivalent single-phase basis.

3. Permanent Faults and Temporary Faults

Faults may be classified as permanent or temporary. Permanentfaults are those in which insulation failure or structure failure produces damage that makes operation of the equipment impossible and requires repairs to be made. Temporary faults are those which may be removed by deenergizing the equipment for a short period of time, short circuits on overhead lines frequently are of this nature. High winds may cause two or more conductors to swing together momentarily. During the short period of contact an arc is formed which may continue as long as the line remains energized. However, if automatic equipment can be brought into operation to deenergize the line quickly, little physical damage may result and the line may be restored to service as soon as the arc is extinguished. Arcs across insulators due to overvoltages from lightning or switching transients usually can be cleared by automatic circuit-breaker operation before significant structure damage occurs.

Because of this characteristic of faults on lines, many companies operate following a procedure known as high-speed reclosing. On the occurrence of a fault, the line is promptly deenergized by opening the circuit breakers at each end of the line. The breakers remain open long enough for the arc to clear, and thenreclose automatically. In many instances service is restored in a fraction of a second. Of course, if structure damage has occurred and the fault persists, it is necessary for the breakers to reopen and lock open.

Words and Phases

intentional *adj*. 有意的，有心的
occurrence *n*. 事件，事故，发生
disconnect from 使分离，切断
sleet *n*. 雨雪，冰凌
oil-impregnated *adj*. 浸过油的
polyethylene *n*. 聚乙烯
deterioration *n*. 恶化，劣化，老化
ionization *n*. 离子化，电离
line to-neutral 从导线到中性点的
deenergize *n*. 断开，断电
permanent fault 永久故障
temporary fault 暂时故障
extinguished *n*. 熄灭
reclosing *n*. 重合闸

Notes

1. It is apparent that the power system designer must anticipate points at which faults may occur，be able to calculate conditions that exist during a fault，and provide equipment properly adjusted to open the switches necessary to disconnect the faulted equipment from the remainder of the system.

显然，系统设计者事先考虑到故障可能发生的地方，能够预测故障发生期间的各种状况，提供调节好的设备，以便驱动为将故障设备切除所必须断开的开关能够跳闸。

2. These materials undergo some deterioration with age，particularly if overloads on the cables have resulted in their operation at elevated temperature.

这些绝缘材料会随着时间的流逝而老化，尤其是在过负荷引起高温下运行的时候更是如此。

3. Any small void present in the body of the insulating material will result in ionization of the gas contained therein，the products of which react unfavorably with the insulation.

绝缘材料内的空隙会导致气体的电离，电离的生成物会对绝缘不利。

4. Major insulation may fail，permitting arcs to be established between primary and secondary windings or between the winding and grounded metal parts such as the core or tank.

线匝之间的绝缘损坏会在变压器的一次绕组和二次绕组之间或绕组与接地金属器件（如铁芯和外壳）之间产生电弧。

Exercises

1. Answer the following questions according to the text

(1) What damages may be caused whenfault is occurred?

(2) What factors can bring the faults of overhead lines?

(3) What factors can bring the faults of underground cables?

(4) What damages may be occurred when the insulation failure between adjacent turns of the same winding?

(5) What are the permanent faults and temporary faults?

2. Translate the following sentences into Chinese according to the text

(1) Faults are the unintentional or intentional connecting together of two or more conductors which ordinarily operate with a difference of potential between them.

(2) Even in the case of less severe faults, high currents in the faulted system may overheat equipment; sustained overheating may reduce the useful life of the equipment.

(3) Overvoltages due to lighting or switching may cause flashover of supporting or from conductor to conductor.

(4) The possibility of cable failure is increased if lighting or switching produces transient voltage of abnormally high values between the conductors.

(5) Temporary faults are those which may be removed by deenergizing the equipment for a short period of time, short circuits on overhead lines frequently are of this nature.

3. Translate the following into Chinese

A distinction must be made between a fault and an overload. An overload implies only that loads greater than the designed values have been imposed on system. Under such a circumstance the voltage at the overload point may be low, but not zero. This under-voltage condition may extend for some distance beyond the overload point into the remainder of the system. The currents in the overloaded equipment are high and may exceed the thermal design limits. Nevertheless, such currents are substantially lower than in the case of a fault. Service frequently may be maintained, but at below-standard voltage.

Overloads are rather common occurrences in homes. To prevent such trouble residential circuits are protected by fuses or circuit breakers which open quickly when currents above specified values persist. Distribution transformers are sometimes overloaded as customers install more and more appliances. The continuous monitoring of distribution circuits is necessary to be certain that transformer sizes are increased as load grows.

Text B　Power System Protection

Good design, maintenance, and proper operating procedures can reduce the probability of occurrence of faults, but cannot eliminate them. Given that faults will inevitably occur, the objective of protective system design is to minimize their impact.

Protection systems have three basic components:

(1) Instrument transformers.

(2) Relays.

(3) Circuit breakers.

Faults are removed from a system by opening or "tripping" circuit breakers. These are the same circuit breakers used in normal system operation for connecting or disconnecting generators, lines and loads. For emergency operation the breakers are tripped automatically when a fault condition is detected. Ideally, the operation is highly selective; only those breakers closest to the fault operate to remove or "clear" the fault. The rest of the system remains intact.

Fault conditions are detected by monitoring voltages and currents at various critical points in the system. Abnormal values individually or in combination cause relays to operate, energizing tripping circuits in the circuit breakers. A simple example is shown in Fig. 4. 10.

Fig. 4. 10 shows a simple overcurrent protection schematic with: ① one type of instrument transformer called the current transformer, ② an overcurrent relay, and ③ a circuit breaker for a single-phase line. The function of the current transformer is to reproduce in its secondary winding a current I' that is proportional to the primary current I. The current transformer converts primary currents in the kiloamp range to secondary currents in the 0 ~ 5 A range for convenience of measurement, with the following advantages.

Safety: Instrument transformers provide electrical isolation from the power system so that personnel working with relays will work in a safe environment.

Economy: Lower-level relay inputs enable relays to be smaller, and less expensive.

Accuracy: Instrument transformers accurately reproduce power currents and voltages over wide operating ranges.

The function of relay is to discriminate between normal operation and fault conditions. The overcurrent relay in Fig. 4. 10 has an operating coil, which is connected to the current transformer secondary winding, and a set of contacts. When $|I'|$ exceeds a specified "pickup" value, the operating coil causes the normally open contacts to close. When the relay contacts close, the trip coil of the circuit breaker is energized, which then cause the circuit breaker to open.

Note that the circuit breaker does not open until its operating energized, either manually or by relay operation. Based on information from instrument transformers, a decision is made and "relayed" to the trip coil of the breaker, which actually opens the power circuit, hence the name relay.

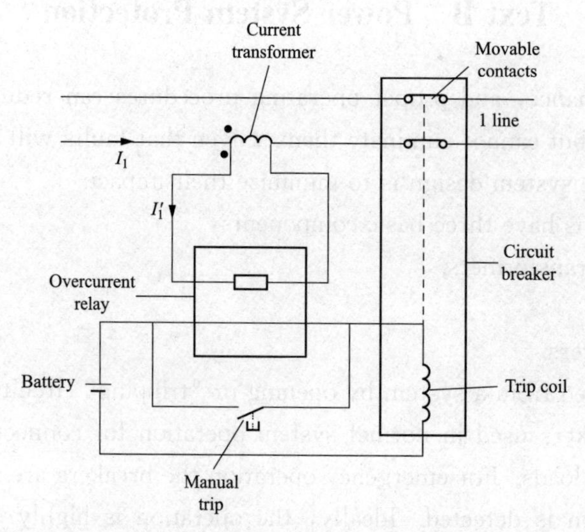

Fig. 4. 10 Schematic of overcurrent protection

In more general cases than the one we have been describing, potential (voltage) transformers as well as current transformers are used. Collectively these transformers are known as instrument transformers. It is note that these transformers are used for measurements (instrumentation) as well as in protective schemes. It is not practical to use potential transformers at the very high line voltages frequently encountered. Instead, a portion of the line voltage is used derived from a voltage-divider circuit composed of series capacitors. The output is then fed through series inductance to a potential transformer, where the voltage is reduced further.

System-protection components have the following design criteria.

(1) Reliability: operate dependably when fault conditions occur, even after remaining idle for months or years. Failure to do so may result in costly damages.

(2) Selectivity: A void unnecessary, false trips.

(3) Speed: Operate rapidly to minimize fault duration and equipment damage. Any intentional time delays should be precise.

(4) Economy: Provide maximum protection at minimum cost.

(5) Simplicity: Minimize protection equipment and circuitry.

Since it is impossible to satisfy all these criteria simultaneously, compromises must be made in system Protection.

Words and Phases

maintenance *n.* 维护，维修

operating procedure 操作规程

instrument transformer 仪表用变压器

tripping *n.* 跳闸，分离

critical *adj.* 临界的

current transformer 电流互感器

primary current 一次电流，原边电流

secondary current 二次电流，副边电流

discriminate *v.* 区分，识别

pick up value 始动值

manually *adv.* 手动地

potential transformer 电压互感器

voltage-divider 分压器

Notes

Instead, a portion of the line voltage is used derived from a voltage-divider circuit composed of series capacitors. The output is then fed through series inductance to a potential transformer, where the voltage is reduced further.

相反，通过分压器（由串联电容器组成）可以获得线电压的部分数值，电压互感器的输出通过与其串联的电感来输出，此时的电压已经被大大地降低了。

Exercises

1. Answer the following questions according to the text

(1) What basic components are protection systems made up of?

(2) What physical parameters are being used to detect the fault conditions?

(3) What is the function of the current transformer?

(4) What instrument transformers are also used besides current transformers?

(5) What are design criteria in the system-protection components?

2. Translate the following sentences into Chinese according to the text

(1) Given that faults will inevitably occur, the objective of protective system design is to minimize their impact.

(2) Fault conditions are detected by monitoring voltages and currents at various critical points in the system.

(3) The overcurrent relay in Fig. 4. 10 has an operating coil, which is connected to the current transformer secondary winding, and a set of contacts.

(4) Note that the circuit breaker does not open until its operating energized, either manually or by relay operation.

(5) In more general cases than the one we have been describing, potential (voltage) transformers as well as current transformers are used.

3. Translate the following into Chinese

The IEEE defines a relay as "a device whose function is to detect defective lines or apparatus or other power system conditions of an abnormal or dangerous nature and to initiate appropriate control action". In practice, a relay is a device that closes or opens a contact when energized. Relays are also use in low-voltage (600 V and below) power systems and almost anywhere that electricity is used. They are used in heating, air conditioning, stoves, clothes washers and dryers, refrigerators, dishwashers, telephone networks, traffic controls, airplane and other transportation systems, and robotics, as well as many other applications.

Problems with the protection equipment itself can occur. A second line of defense, called backup relays, may be used to protect the first line of defense, called primary relays. In HV and EHV systems, separate current or voltagemeasuring devices, separate trip coils on the circuit breakers, and separate batteries for the trip coils may be used. Also, the various protective devices must be properly coordinated such that primary relays assigned to protect equipment in a particular zone operate first. If the primary relays fail, then backup relays should operate after a specified time delay.

Unit 6 Power System Control

Text A Introduction of Power System Control

Automatic control systems are used extensively in power systems. Local controls are employed at turbinegenerator units and at selected voltage-controlled buses. Central controls are employed at area control centers.

Fig. 4. 11 shows two basic controls of a steam turbinegenerator: the voltage regulator and turbinegovernor. The voltage regulator adjusts the power output of the generator exciter in order to control the magnitude of generator terminal voltage U_t. When a reference voltage U_{ref} is raised (or lowered), the output voltage U_r of the regulator increases (or decreases) the exciter voltage E_{fd} applied to the generator field winding, which in turn acts to increase (or decrease) U_t. Also a voltage transformer and rectifier monitor U_t, which is used as a feedback signal in the voltage regulator. If U_t decreases, the voltage regulator increases U_r to increase E_{fd}, which in turn acts to increase U_t.

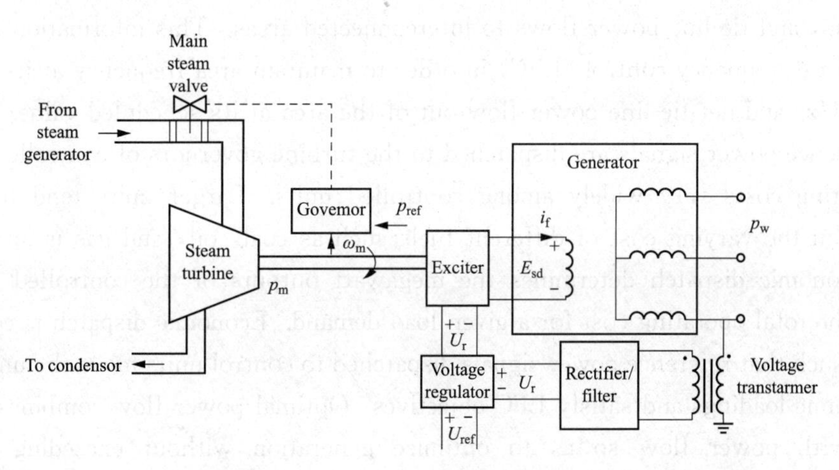

Fig. 4. 11 Voltage regulator and turbine-governor controls for a steam-turbine generator

In addition to voltage regulators at generator buses, equipment is used to control voltage magnitudes at other selected buses. Tap-changing transformers, switched capacitor banks, and static var systems can be automatically regulated for rapid voltage control.

Central controls also play an important role in modern power systems. Today's systems are composed of interconnected areas, where each area has its own control center. There are many advantages to interconnections. For example, interconnected areas can share their reserve power to handle anticipated load peaks and unanticipated generator outages. Interconnected areas can also tolerate larger load changes with smaller frequency deviations than an isolated area.

Fig. 4. 12 shows how a typical area meets its daily load cycle. The base load is carried by based-loaded generators running at 100% of their rating for 24 hours. Nuclear units and large fossil-fuel units are typically base-loaded. The variable part of the load is carried by units that are controlled from the central control center. Medium-sized fossil-fuel units and hydro units are used for control. During peak load hours, smaller, less efficient units such as gasturbine or diesel-generating units are employed. In addition, generators provide a reserve margin.

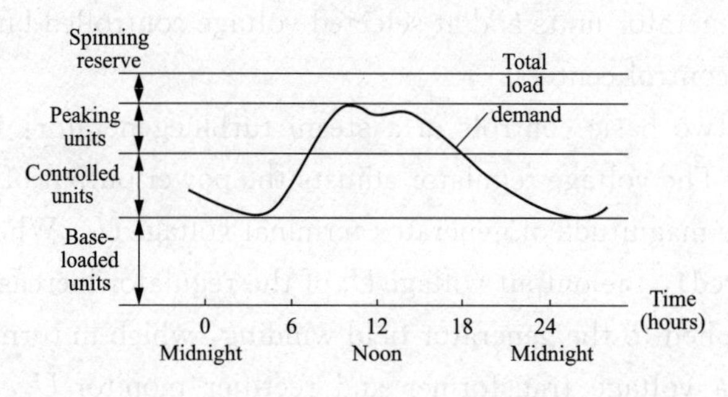

Fig. 4. 12 Daily load cycle

The central control center monitors information including area frequency, generating unit outputs, and tie-line power flows to interconnected areas. This information is used by automatic load-frequency control (LFC) in order to maintain area frequency at its scheduled value (60 Hz) and net tie-line power flow out of the area at its scheduled value. Raise and lower reference power signals are dispatched to the turbine-governors of controlled units.

Operating costs vary widely among controlled units. Larger units tend to be more efficient, but the varying cost of different fuels such as coal, oil, and gas is an important factor. Economic dispatch determines the megawatt outputs of the controlled units that minimize the total operating cost for a given load demand. Economic dispatch is coordinated with LFC such that reference power signals dispatched to control units move the units toward their economic loadings and satisfy LFC objectives. Optimal power flow combines economic dispatch with power flow so as to optimize generation without exceeding limits on transmission line load ability.

Words and Phases

local control 局部控制

central control 中央控制，集中控制

employ *v.* 采用，使用

turbine governor *n.* 涡轮调速器

terminal voltage 端电压

reference voltage 基准电压，参考电压

tap-changing *n. adj.* 抽头转换，切换抽头的

capacitor banks 电容器组

base load 基本负荷

peak load 峰值负荷，高峰负荷

scheduled value 预定的数值，计划数值

dispatch to 派遣，发送

Notes

1. When a reference voltage U_{ref} is raised (or lowered), the output voltage U_r of the regulator increases (or decreases) the exciter voltage E_{fd} applied to the generator field winding, which in turn acts to increase (or decrease) U_t.

当参考电压升高或降低时，调解器的输出电压 U_r 增加以增加发电机励磁绕组的端电压，励磁电压的增加又会使发电机端电压 U_t 增加。

2. Optimal power flow combines economic dispatch with power flow so as to optimize generation without exceeding limits on transmission line load ability.

最优功率流控制与功率流的经济调度相结合，从而使得发电得到很好控制而不至于超过传输线负荷能力的上限。

Exercises

1. Answer the following questions according to the text

(1) Where are local controls in power systems employed?

(2) Where are central controls in power systems employed?

(3) What equipment are used to control voltage magnitudes besides voltage regulators?

(4) What carries the base load of a typical area?

(5) Which information is being monitored by the central control centers?

2. Translate the following sentences into Chinese according to the text

(1) The voltage regulator adjusts the power output of the generator exciter in order to control the magnitude of generator terminal voltage U_t.

(2) Tap-changing transformers, switched capacitor banks, and static var systems can be automatically regulated for rapid voltage control.

(3) During peak load hours, smaller, less efficient units such as gasturbine or diesel-generating units are employed.

(4) Larger units tend to be more efficient, but the varying cost of different fuels such as coal, oil, and gas is an important factor.

(5) Economic dispatch is coordinated with LFC such that reference power signals dispatched to control units move the units toward their economic loadings and satisfy LFC objectives.

3. Translate the following sentences into Chinese

The exciter delivers DC power to the field winding on the rotor of a synchronous generator. For older generators, the exciter consists of a DC generator driven by the rotor. The DC power is transferred to the rotor via ship rings and brushes. For newer generators, static or brushless exciters are often employed.

For static exciters, AC power is obtained directly from the generator terminals or a nearby station service bus. The AC power is then rectified via thyristors and transferred to the rotor of the synchronous generator via slip rings and brushes.

For brushless exciters, AC power is obtained from an "inverted" synchronous generator whose three-phase armature windings are located on the main generator rotor and whose field winding is located on the stator. The AC power from the armature windings is rectified via diodes mounted on the rotor and is transferred directly to the field winding. For this design, slip rings and brushes are eliminated.

Text B Controlling the Power Balance between Generator and Load

1. Controlling the Power Balance between Generator and Load

The electrical energy consumed by the thousands of customers must immediately be supplied by the AC generators because electrical energy cannot be stored. How do we maintain this almost instantaneous balance between customer requirements and generated power? To answer the question, let us consider a single hydropower station supplying a regional load R_1 (Fig. 4. 13). Water behind the dam flows through the turbine, causing the turbine and generator to rotate.

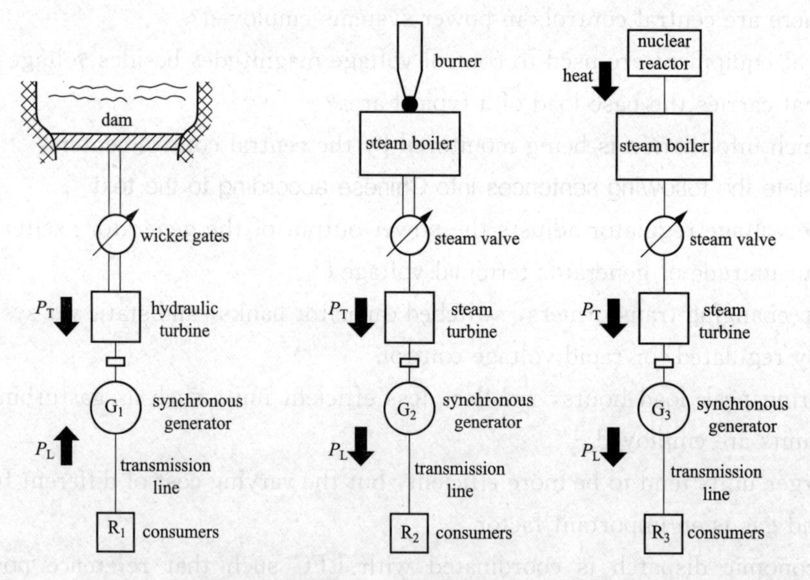

Fig. 4. 13 Power supplied to three regions

The mechanical power P_T developed by the turbine depends exclusively on the opening of the wicket gates that control the water flow. The greater the opening, the more water is admitted to the turbine and the increased power is immediately transmitted to the generator.

On the other hand, the electric power P_L, drawn from the generator depends exclusively on the load. When the mechanical power P_T supplied to the rotor is equal to the electrical power P_L consumed by the load, the generator is in dynamic equilibrium and its speed remains constant. The electrical system is said to be stable.

However, we have just seen that the system demand fluctuates continually, so P_L is sometimes greater and sometimes less than P_T. If P_L is greater than P_T, the generating unit (turbine and generator) begins to slow down. Conversely, if P_L is less than P_T, the generating unit speeds up.

The speed variation of the generator is, therefore, an excellent indicator of the state of equilibrium between P_L and P_T, hence, of the stability of the system. If the speed falls the wicket gates must open, and it rises they must close so as to maintain a continuous state of equilibrium between P_L and P_T. Although we could adjust the gates manually by observing the speed, an automatic speed regulator is always used.

Speed regulators, or governors, are extremely sensitive devices. They can detect speed changes as small as 0.02 percent. Thus, if the speed of a generator increases from 1800 r/min to 1800.36 r/min, the governor begins to act on the wicket gate mechanism. If the load should suddenly increase, the speed will drop momentarily, but the governor will quickly bring it back to rated speed. The same corrective action takes place when the load is suddenly removed.

Clearly, any speed change produces a corresponding change in the system frequency. The frequency is therefore an excellent indicator of the stability of a system. The system is stable so long as the frequency is constant.

The governors of thermal and nuclear stations operate the same way, except that they regulate the steam valves, allowing more or less steam to flow through the turbines. The resulting change in steam flow has to be accompanied by a change in the rate of combustion. Thus, in the case of a coal-burning boiler, we have to reduce combustion as soon as the valves are closed off, otherwise the boiler pressure will quickly exceed the safety limits.

2. Advantage of Interconnected Systems

Consider the three generating stations of Fig. 4.14, connected to their respective regional loads R_1, R_2, and R_3. Because the three systems are not connected, each can operate at its own frequency, and a disturbance on one does not affect the others. However, it is preferable to interconnect the systems because: ① it improves the overall stability, ② it provides better continuity of service, and ③ it is more economical. Fig. 4.14 shows four interconnecting transmission lines, tying together both the generating stations and the regions being serviced. High-speed circuit breakers d_1 to d_{10} are installed to automatically interrupt power in case of a fault and to reroute the flow of electric power. We now discuss the advantages of such a network.

(1) Stability. Systems that are interconnected have greater reserve power than a system working alone. In effect, a large system is better able to withstand a large disturbance and, consequently, it is inherently more stable. For example, if the load suddenly increases in region R_1, energy immediately flows from stations G_2 and G_3 and over the interconnecting tie-lines. The heavy load is, therefore, shared by all three stations instead of being carried by one alone.

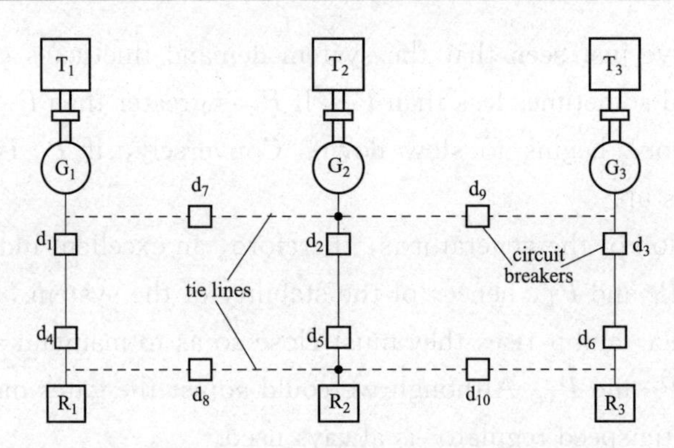

Fig. 4. 14 Three network connected by four tie-lines

(2) Continuity of Service. If a generating station should break down, or if it has to be shut down for annual inspection and repair, the customers it serves can temporarily be supplied by the two remaining stations. Energy flowing over the tie-lines is automatically metered and credited to the station that supplies it, less any wheeling charges. A wheeling charge is the amount paid to another electric utility when its transmission lines are used to deliver power to a third party.

(3) Economy. When several regions are interconnected, the load can be shared among the various generating stations so that the overall operating cost is minimized. For example, instead of operating all three stations at reduced capacity during the night when demand is low, we can shut down one station completely and let the other carry the load. In this way we greatly reduce the operating cost of one station while improving the efficiency of the other stations, because they now run closer to their rated capacity.

Electric utility companies are, therefore, interested in grouping their resources by a grid of inter-connecting transmission lines. A central dispatching office (control center) distributes the load among the various companies and generating stations so as to minimize the costs. Due to the complexity of some systems, control decisions are invariably made with the aid of a computer. The dispatching office also has to predict daily and seasonal load changes and to direct the start-up and shut-down of generating units so as to maintain good stability of the immense and complicated network.

Although such interconnected systems must necessarily operate at the same frequency, the load can still be allocated among the individual generating units, according to a specific program. Thus, if a generating unit has to deliver more power, its governor setting is changed slightly so that more power is delivered to the generator. The increased electrical output from this unit produces a corresponding decrease in the total power supplied by all the other generating units of the interconnected system.

Words and Phases

wicket *n.* 闸门 dynamic equilibrium 动态平衡

steam valves 蒸汽阀
coal-burning boiler 燃煤锅炉
overall stability 总稳定性，总稳定度
high-speed circuit breaker 高速断路器
reserve power 备用容量，备用功率
withstand *v.* 抵抗，经受住

Notes

1. The greater the opening, the more water is admitted to the turbine and the increased power is immediately transmitted to the generator.

闸门打开的越大就会有更多的水流注入涡轮，进而发电机就会产生更大的电能。

2. The dispatching office also has to predict daily and seasonal load changes and to direct the start-up and shut-down of generating units so as to maintain good stability of the immense and complicated network.

调度中心也不得不预测负荷的日变化和季度变化，从而来直接决定启用或关闭某些发电机组以保证大电网有一个好的稳定性。

Exercises

1. Answer the following questions according to the text

(1) How do we maintain the instantaneous balance between customer requirements and generated power?

(2) What is the dynamic equilibrium of generator?

(3) In order to maintain a continuous state of equilibrium between PL and PT, the wicket gates should be opened or closed?

(4) What are the advantages of the interconnected systems?

(5) Why the interconnected system is more economical?

2. Translate the following sentences into Chinese according to the text

(1) The electrical energy consumed by the thousands of customers must immediately be supplied by the AC generators because electrical energy cannot be stored.

(2) The speed variation of the generator is, therefore, an excellent indicator of the state of equilibrium between PL and PT, hence, of the stability of the system.

(3) If the load should suddenly increase, the speed will drop momentarily, but the governor will quickly bring it back to rated speed.

(4) High-speed circuit breakers d_1 to d_{10} are installed to automatically interrupt power in case of a fault and to reroute the flow of electric power.

(5) A central dispatching office (control center) distributes the load among the various companies and generating stations so as to minimize the costs.

3. Translate the following sentences into Chinese

As electric utilities have grown in size and the number of interconnections has increased, planning for future expansion has become increasingly complex. The increasing cost of additions and modifications has made it imperative that utilities consider a range of design options, and perform detailed studies of the effects on the system of each option, based on a number of assumptions: normal and abnormal operating conditions, peak and off-peak loadings, and present and future years of operation. A large volume of network data must also be collected and accurately handled. To assist the engineer in this power system planning, digital computers and highly developed computer programs are used. Such programs include powerflow, stability, short-circuit, and transients programs.

Unit 7 Modern Power Systems

Text A Distributed Generations

Conventional power stations, such as coal-fired, gas and nuclear powered plants, as well as hydroelectric dams and large-scale solar power stations, are centralized and often require electricity to be transmitted over long distances. By contrast, distributed generations (DGs), generated or stored by a variety of small, grid-connected devices referred to as distributed energy resources (DER), are decentralized, modular and more flexible technologies, that are located close to the load they serve, albeit having capacities of only 10 MW or less. These systems can comprise multiple generation and storage components. In this instance they are referred to as Hybrid power system. DER systems typically use renewable energy sources, including small hydro, biomass, biogas, solar power, wind power, and geothermal power, and increasingly play an important role for the electric power distribution system.

The primary drivers behind the growth of DG and the current focus on its integration into electric power system operation and planning can be classified into three main categories, namely environmental, commercial and national/regulatory. These drivers are discussed briefly below.

1. Environmental drivers

The use of renewable energy and combined heat and power (CHP) to limit greenhouse gas emissions is one of the main drivers for DG. In this regard, it is important to point out that integration of renewable sources of electrical energy into power systems is a somewhat different question from that of the integration of DG into power systems. Integration of DG includes some of the issues related to integration of renewable sources but clearly does not deal with integration of transmission connected renewable sources such as large on shore and off shore wind farms.

Another important driver for DG from the environmental perspective is the avoidance of construction of new transmission lines and large power plants to which there is increasing public opposition. There is however also opposition from some environmental lobby groups to onshore wind farms on grounds of noise and visual "pollution". There is therefore a balance to be struck between the need for sustainable energy solutions on the one hand and the need to maintain scenic beauty of the environment. Some argue that environmentally benign technologies, such as wind, that do not emit any greenhouse gas and have no long term waste management problems should be favored. Technological developments in generator technology are already delivering cost effective small to medium size generation technologies for domestic application such as micro-CHP. In times when a premium is placed on land use these technologies are likely to prove popular.

2. Commercial drivers

One of the acknowledged consequences of the introduction of competition and choice in electricity is the increased risk faced by all players in the electricity supply chain from generators through transmission and distribution businesses to retailers. It is well known that the capital outlay required to establish new power stations can be very high. The uncertainties associated with a competitive market environment may favor generation projects with a small capacity whose financial risk is commensurately small.

The presence of DG close to load centers can have a beneficial impact on power quality and supply reliability. One area of improvement is voltage profile improvements, reduction in number of customer minutes lost especially if DG is allowed and able to stay on when there are network outages (islanding).

3. National/regulatory drivers

In recent times, there has been increasing concern amongst energy policy makers regarding energy security. There is a recognition that modern societies have become so dependent on energy resources to the extent that should there be a disruption in its supply the consequences would be too ghastly to contemplate in political, economic and social terms. Because of this the EU energy policy focuses on energy security and sustainability.

In the context of energy security and sustainability, DG is an attractive proposition in many respects. Some of the more important ones are for example:

It is distributed around the network close to customers—failure of one power station will have limited impact on the whole system compared to failure of one large power plant or bulk electricity transmission facility.

Diverse technologies and primary energy sources—by diversifying the energy sources especially utilizing renewable sources there is sense of control over the nation's future energy needs. There is increasing concern that the bulk of fossil fuel based energy supplies come from regions of the world where control of these resources could be potentially unpredictable thus posing an unacceptable risk.

There is an abiding faith amongst the proponents of reform of electricity supply industries that introduction of competition in generation and customer choice will deliver low energy prices and better service quality. One of the prerequisites for effective competition to occur is that there must be many players in the market. DG clearly advances this cause by providing many small generators that could potentially trade in the energy market and, where appropriate market arrangements exist, also trade in ancillary services.

Words and Phases

hydroelectric dam 水电大坝	distributed generation 分布式电源
solar power station 太阳能电厂	decentralized *adj*. 分散的

renewable *adj*. 可再生的
biomass *n*. 生物质
geothermal *adj*. 地热的
environmental *adj*. 环境的
commercial *adj*. 商业的
combined heat and power 热电联产
greenhouse gas 温室气体
sustainable *adj*. 可持续的
retailer *n*. 零售商
commensurately *adv*. 相当地
reliability *n*. 可靠性
energy policy 能源政策
energy security and sustainability 能源安全和可持续性
bulk electricity transmission facility 大容量输电设备
diversify *v*. 多样化
unpredictable 不可预测的
electricity supply 电力供应
prerequisite *n*. 先决条件
ancillary service 辅助服务

Notes

1. DER systems typically use renewable energy sources, including small hydro, biomass, biogas, solar power, wind power, and geothermal power, and increasingly play an important role for the electric power distribution system.

分布式能源系统通常使用可再生能源，例如小型水电站、生物质能、沼气、太阳能、风能和地热能，分布式能源系统在电力配电系统中扮演着越来越重要的角色。

2. The use of renewable energy and combined heat and power (CHP) to limit greenhouse gas emissions is one of the main drivers for DG.

利用可再生能源和热电联产来限制温室气体排放是分布式发电的主要驱动因素之一。

3. There is an abiding faith amongst the proponents of reform of electricity supply industries that introduction of competition in generation and customer choice will deliver low energy prices and better service quality.

供应行业改革的支持者们坚信，在发电和客户选择方面引入竞争将带来更低的能源价格和更好的服务质量。

Exercises

1. Answer the following questions according to the text

(1) What is the significant difference between conventional power stations and DGs?

(2) What are the features of DERs?

(3) What kinds of energy do DER systems typically use?

(4) What are the advantages and disadvantages of the introduction of competition and choice in electricity, and why?

(5) In the context of energy security and sustainability, DG is an attractive proposition in many respects. Explain the reasons.

2. Translate the following sentences into Chinese according to the text

(1) Another important driver for DG from the environmental perspective is the avoidance of construction of new transmission lines and large power plants to which there is increasing public opposition.

(2) The uncertainties associated with a competitive market environment may favour generation projects with a small capacity whose financial risk is commensurately small.

(3) There is a recognition that modern societies have become so dependent on energy resources to the extent that should there be a disruption in its supply the consequences would be too ghastly to contemplate in political, economic and social terms.

(4) DG clearly advances this cause by providing many small generators that could potentially trade in the energy market and, where appropriate market arrangements exist, also trade in ancillary services.

3. Translate the following paragraph into Chinese

Nowadays it is more common for DG to be considered in the context of the wider concept of distributed energy resources (DER), which includes not only DG but also energy storage and responsive loads. The power system architecture of the future, incorporating DER, will look very different from that of today. Whilst the pace of change is likely to be evolutionary, the change itself is expected to be nothing short of a revolution as many traditionally held views and approaches to system operation and planning developed over the last 100 years are challenged and transformed to suit the requirements envisaged in the brave new world of the future.

Text B Active Distribution Network

Active distribution network management is seen as the key to cost effective integration of DG into distribution network planning and operation. This is in direct contrast to the current connect and forget approach.

The historic function of "passive" distribution networks is viewed primarily as the delivery of bulk power form the transmission network to consumers at lower voltages. These networks were designed through deterministic (load flow) studies considering the critical cases so that distribution networks could operate with a minimum amount of control. In other words control problems were solved at the planning stage. This practice of passive operation can limit the capacity of distributed generation that can be connected to an existing system.

For well-designed distribution circuits, there is little scope for distributed generation when simple deterministic rules (e. g. consideration of minimum load and maximum generation) are used. This practice significantly limits the connection of DG. As these conditions may only apply for a few hours per year it is clearly desirable to consider stochastic voltage limits, as proposed under European standard EN 50160. The application of probabilistic load flow and Monte-Carlo simulation techniques provide probabilities of voltage limit violations and thus leads to objective decision-making. In Fig. 4. 15, example of a probabilistic load flow analysis is provided showing the probability density Fig. 4. 15. Impact of wind farm output on the frequency distribution of voltages at node. Functions of the voltage at the connection bus of a wind park for low levels of installed capacity. For low levels of penetration (0. 675 MW), the voltage was generally below 1. 0 per unit. However, at higher levels of penetration (8. 0 MW), the likelihood of high voltages increased. With a high penetration, the voltage rise effect is very clear (15% above nominal). It is, however, important to note that whilst the voltage rise is pronounced, the probability of its occurrence is low, less than 6% in this particular case. Using crude studies, connecting a larger wind farm (8 MW) may have been rejected without considering the likelihood of the increase in voltage. No attention would have been paid to the fact that the mean voltage level over time would be relatively unaffected by the size of the wind farm.

This analysis would also allow distributed generation to decide to be constrained off in certain circumstance to limit voltage rise. Further, many DGs have the ability to operate at various power factors and may even be able to act as sources/sinks of reactive power when not generating. For some overhead distribution circuits (i. e. those with high reactance) then the DGs could contribute to circuit voltage control provided suitable control and commercial systems were in place.

In contrast, active management (AM) techniques enable the distribution network operator to maximize the use of the existing circuits by taking full advantage of generator dispatch, control of transformer taps, voltage regulators, reactive power management and system reconfiguration in an integrated manner. AM of distribution networks can contribute to the balancing of generation with load and ancillary services. In future, distribution management systems could provide real-time network monitoring and control at key network nodes by communicating with generator controls, loads and controllable network devices, such as reactive compensators, voltage regulators and on load tap changing-transformers (OLTC). State estimation and real-time modelling of power capability, load flow, voltage, fault levels and security could be used to make the right scheduling/constraining decisions across the network. These techniques will probably be applied gradually rather than fulfilling all the above listed attributes right from the beginning (Bopp et al. , 2003).

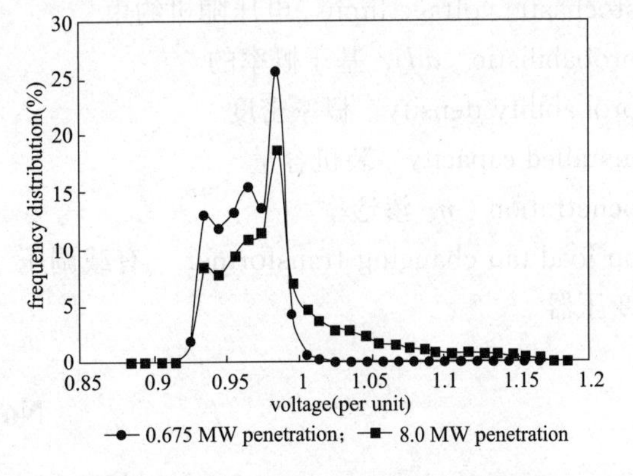

Fig. 4. 15 Impact of wind farm output on the frequency distribution of voltages at node

The DMS controller software has two functional blocks (see Fig. 4. 16): state estimation and control scheduling. The state estimation block uses the network electrical parameters, network topology, load models and real-time measurements to calculate a network state estimate. The measurement input comprises the local and network measurements. This is passed to the control scheduling block, which uses it to calculate a new set of control values for the devices connected to the network. The set of control values optimizes the power flow in the network, whilst observing all the constraints and taking account of all the contracts.

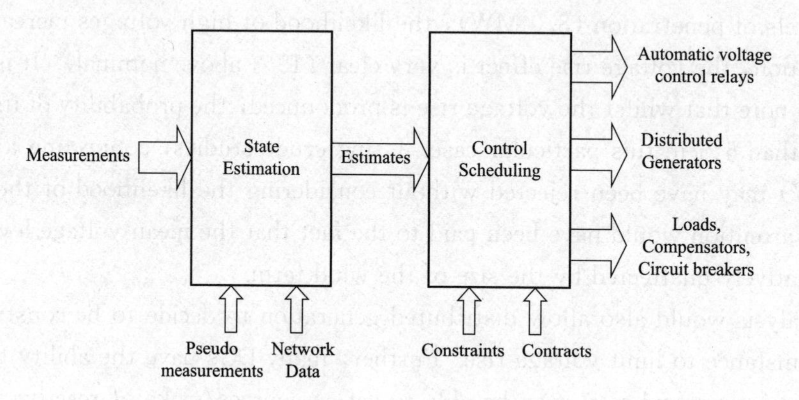

Fig. 4. 16 Block diagram of DMS controller software

Words and Phases

active distribution network 主动配电网
cost effective 成本效益
planning and operation 规划与运行
deterministic *adj*. 决定性的
stochastic voltage limit 电压随机约束
probabilistic *adj*. 基于概率的
probability density 概率密度
installed capacity 装机容量
penetration *n*. 渗透率
on load tap changing-transformer 有载调压变压器
distribution network operator 电力系统运营商
dispatch *n*. 调度
reconfiguration *n*. 重构
real-time network monitoring 实时网络监控
state estimation 状态估计
network topology 网络拓扑
constraint *n*. 约束
contract *n*. 合同

Notes

1. Active distribution network management is seen as the key to cost effective integration of DG into distribution network planning and operation.

配电系统的主动管理被视为实现配电网规划与运营的成本效益整合的关键。

2. The historic function of "passive" distribution networks is viewed primarily as the delivery of bulk power form the transmission network to consumers at lower voltages.

被动式配电网的传统功能主要被视为向低电压处的用户输送大量电力

3. Further, many DGs have the ability to operate at various power factors and may even be able to act as sources/sinks of reactive power when not generating.

此外，许多 DGs 具有在各种功率因数下运行的能力，甚至可以在不发电时充当无功功率源。

Exercises

1. Answer the following questions according to the text

(1) Why is active distribution network management necessary?

(2) What can significantly limit the connection of DG?

(3) How do DGs influence the voltage profile of active distribution networks?

(4) What kind of techniques does active management (AM) include for the distribution network operator to maximize the use of the existing circuits?

(5) The DMS controller software has two functional blocks: state estimation and control scheduling. Please explain the functions of the two parts.

2. Translate the following sentences into Chinese according to the text

(1) In other words control problems were solved at the planning stage. This practice of passive operation can limit the capacity of distributed generation that can be connected to an existing system.

(2) No attention would have been paid to the fact that the mean voltage level over time would be relatively unaffected by the size of the wind farm.

(3) This is passed to the control scheduling block, which uses it to calculate a new set of control values for the devices connected to the network.

(4) In contrast, active management (AM) techniques enable the distribution network operator to maximize the use of the existing circuits by taking full advantage of generator dispatch, control of transformer taps, voltage regulators, reactive power management and system reconfiguration in an integrated manner.

3. Translate the following paragraph into Chinese

Coordinated voltage control with on load tap changers and voltage regulators generally leads to operating arrangements requiring a number of measurements from key network points as well as communications. From a regulator's perspective, active management should enable open access to distribution networks. It has the function of facilitating competition and the growth of small-scale generation. In addition, the use of the existing distribution assets should be maximized to minimize costs to consumers. Therefore, an integral understanding of the interrelated technical, economic and regulatory issues of active management and DG is important for the development of the future distribution systems.

Part Ⅴ High Voltage and Insulation Technology

Unit 1 Basic of High Voltage and Insulation Technology

Text A Introduction of Electrical Insulation

In modern times, high voltages are used for a wide variety of applications covering the power systems, industry, and research laboratories. Such applications have become essential to sustain modern civilization. High voltages are applied in laboratories in nuclear research, in particle accelerators, and Van de Graaff generators. For transmission of large bulks of power over long distances, high voltages are indispensable. Also, voltages up to 100 kV are used in electrostatic precipitators, in automobile ignition coils, etc. X-ray equipment for medical and industrial applications also uses high voltages. Modern high voltage test laboratories employ voltages up to 6 MV or more. The diverse conditions under which a high voltage apparatus is used necessitate careful design of its insulation and the electrostatic field profiles. The principal media of insulation used are gases, vacuum, solid, and liquid, or a combination of these. For achieving reliability and economy, knowledge of the causes of deterioration is essential, and the tendency to increase the voltage stress for optimum design calls for judicious selection of insulation in relation to the dielectric strength, corona discharges, and other relevant factors.

1. Electric Field Stresses

Like in mechanical designs where the criterion for design depends on the mechanical strength of the materials and the stresses that are generated during their operation, in high voltage applications, the dielectric strength of insulating materials and the electric field stresses developed in them when subjected to high voltages are the important factors in high voltage systems. In a high voltage apparatus the important materials used are conductors and insulators. While the conductors carry the current, the insulators prevent the flow of currents in undesired paths. The electric stress to which an insulating material is subjected to is numerically equal to the voltage gradient, and is equal to the electric field intensity,

$$E = -\nabla\varphi$$

where E is the electric field intensity, φ is the applied voltage, and operator ∇ is defined as

$$\nabla = a_x \frac{\partial}{\partial x} + a_y \frac{\partial}{\partial y} + a_z \frac{\partial}{\partial z}$$

where a_x, a_y, and a_z are components of position vector $r = a_x x + a_y y + a_z z$.

As already mentioned, the most important material used in a high voltage apparatus is the insulation. The dielectric strength of an insulating material can be defined as the maximum dielectric stress which the material can withstand. It can also be defined as the voltage at which the current starts increasing to very high values unless controlled by the external impedance of the circuit. The electric breakdown strength of insulating materials depends on a variety of parameters, such as pressure, temperature, humidity, nature of applied voltage, imperfections in dielectric materials, material of electrodes, and surface conditions of electrodes, etc. An understanding of the failure of the insulation will be possible by the study of the possible mechanisms by which the failure can occur.

The most common cause of insulation failure is the presence of discharges either within the voids in the insulation or over the surface of the insulation. The probability of failure will be greatly reduced if such discharges could be eliminated at the normal working voltage. Then, failure can occur as a result of thermal or electrochemical deterioration of the insulation.

2. Surge Voltage, Their Distribution and Control

The design of power apparatus particularly at high voltages is governed by their transient behavior. The transient high voltages or surge voltages originate in power systems due to lightning and switching operations. The effect of the surge voltages is severe in all power apparatuses. The response of a power apparatus to the impulse or surge voltage depends on the capacitances between the coils of windings and between the different phase windings of the multi-phase machines. The transient voltage distribution in the windings as a whole are generally very non-uniform and are complicated by traveling wave voltage oscillations set up within the windings. In the actual design of an apparatus, it is, of course, necessary to consider the maximum voltage differences occurring, in each region, at any instant of time after the application of an impulse, and to take into account their durations especially when they are less than one microsecond.

An experimental assessment of the dielectric strength of insulation against the power frequency voltages and surge voltages, on samples of basic materials, on more or less complex assemblies, or on complete equipment must involve high voltage testing. Since the design of an electrical apparatus is based on the dielectric strength, the design cannot be completely relied upon, unless experimentally tested. High voltage testing is done by generating the voltages and measuring them in a laboratory.

When high voltage testing is done on component parts, elaborate insulation assemblies, and complete full-scale prototype apparatus, it is possible to build up a considerable stock of design information; although expensive, such data can be very useful. However, such data can never really be complete to cover all future designs and necessitates use of large factors of safety. A different approach to the problem is the exact calculation of dielectric strength of any insulation arrangement. In an ideal design each part of the dielectric would be uniformly stressed at the maximum value which it will safely withstand. Such an ideal condition is impossible to achieve in practice, for dielectrics of different electrical strengths, due to the practical limitations of construction. Nevertheless it provides information on stress concentration factors—the ratios of maximum local voltage gradients to the mean value in the adjacent regions of relatively uniform stress. A survey of typical power apparatus designs suggests that factors ranging from 2 to 5 can occur in practice; when this factor is high, considerable quantities of insulation must be used.

Words and Phases

particle accelerator 粒子加速器
Van de Graaff generator 范德格拉夫发电机
transmission *n.* 传输，输电
electrostatic precipitator 电气除尘器，静电滤尘器
profile *n.* 剖面
judicious *adj.* 明智的，有见识的，审慎的
corona *n.* 电晕，电晕放电
insulating *adj.* 绝缘的
insulator *n.* 绝缘子，绝缘体
gradient *n.* 梯度，坡度
electric breakdown strength 击穿电压
humidity *n.* 湿度
nature *n.* 种类，类型
imperfection *n.* 缺陷，不完整性
surge *n.* 波动，振荡
originate *vi.* 引起，发生
oscillation *n.* 摆动，振动
power frequency 工频
stress concentration 应力集中

Notes

1. For achieving reliability and economy, knowledge of the causes of deterioration is essential, and the tendency to increase the voltage stress for optimum design calls for judicious selection of insulation in relation to the dielectric strength, corona discharges, and other relevant factors.

为了获得可靠性和节约成本，掌握绝缘劣化的知识是必需的。增加电压应力优化设计的趋势要求明智的选择绝缘，应考虑到绝缘强度，电晕放电和其他相关的因素。

2. Like in mechanical designs where the criterion for design depends on the mechanical strength of the materials and the stresses that are generated during their operation, in high voltage applications, the dielectric strength of insulating materials and the electric field stresses developed in them when subjected to high voltages are the important factors in high voltage systems.

如同机械设计中设计标准依赖于材料在运作时产生的机械强度和应力，高电压应用中，绝缘材料表现出的绝缘强度和电场应力是高电压系统中重要的因素。

Exercises

1. Answer the following questions according to the text

(1) What are the principal media of insulation used in high voltage apparatus?

(2) What is difference between conductors and insulators in a high voltage apparatus?

(3) How to define the dielectric strength of an insulating material?

(4) What parameters depends on the electric breakdown strength of insulating materials?

(5) What is the most common cause of insulation failure?

2. Translate the following sentences into Chinese according to the text

(1) The diverse conditions under which a high voltage apparatus is used necessitate careful design of its insulation and the electrostatic field profiles.

(2) An understanding of the failure of the insulation will be possible by the study of the possible mechanisms by which the failure can occur.

(3) The response of a power apparatus to the impulse or surge voltage depends on the capacitances between the coils of windings and between the different phase windings of the multi-phase machines.

(4) Since the design of an electrical apparatus is based on the dielectric strength, the design cannot be completely relied upon, unless experimentally tested.

(5) However, such data can never really be complete to cover all future designs and necessitates use of large factors of safety.

3. Translate the following paragraph into Chinese

Dielectrics are materials that are used primarily to isolate components electrically from each other or ground or to act as capacitive elements in devices, circuits, and systems. Their insulating properties are directly attributable to their large energy gap between the highest filled valence band and the conduction band. The number of electrons in the conduction band is extremely low, because the energy gap of a dielectric is sufficiently large to maintain most of the electrons trapped in the lower band.

Text B Estimation of Electric Stress

The electric field distribution is governed by the Poisson's equation

$$\nabla^2 \varphi = -\frac{\rho}{\varepsilon_0}$$

where φ is the potential at a given point, ρ is the space charge density in the region, and ε_0 is the electric permittivity of free space (vacuum). However, in most of the high voltage apparatus, space charges are not normally present, and hence the potential distribution is governed by the Laplace's equation

$$\nabla^2 \varphi = 0$$

The operator ∇^2 is called the Laplacian and is a scalar with properties

$$\nabla \cdot \nabla = \nabla^2 = \frac{\partial^2}{\partial x^2} + \frac{\partial^2}{\partial y^2} + \frac{\partial^2}{\partial z^2}$$

There are many methods available for determining the potential distribution, the most commonly used methods being:

(1) The electrolytic tank method.

(2) The method using digital computers.

The potential distribution can also be calculated directly. However, this is very difficult except for simple geometries. In many practical cases, a good understanding of the problem is possible by using some simple rules to sketch the field lines and equipotential. The important rules are:

(1) The equipotential cut the field lines at right angles.

(2) When the equipotential and field lines are drawn to form curvilinear squares, the density of the field lines is an indication of the electric stress in a given region.

(3) In any region, the maximum electric field is given by du/dx, where du is the voltage difference between two successive equipotential dx apart.

Considerable amount of labor and time can be saved by properly choosing the planes of symmetry and shaping the electrodes accordingly. Once the voltage distribution of a given geometry is established, it is easy to redesign the electrodes to minimize the stresses so that the onset of corona is prevented. This is a case normally encountered in high voltage electrodes of the bushings, standard capacitors, etc. When two dielectrics of widely different permittivities are in a series, the electric stress is very much higher in the medium of lower permittivity. Considering a solid insulation in a gas medium, the stress in the gas becomes ε_r times that in the solid dielectric, where ε_r is the relative permittivity of the solid dielectric. This enhanced stress occurs at the electrode edges and one method of overcoming this is to increase the electrode diameter. Other methods of stress control are shown in Fig. 5. 1.

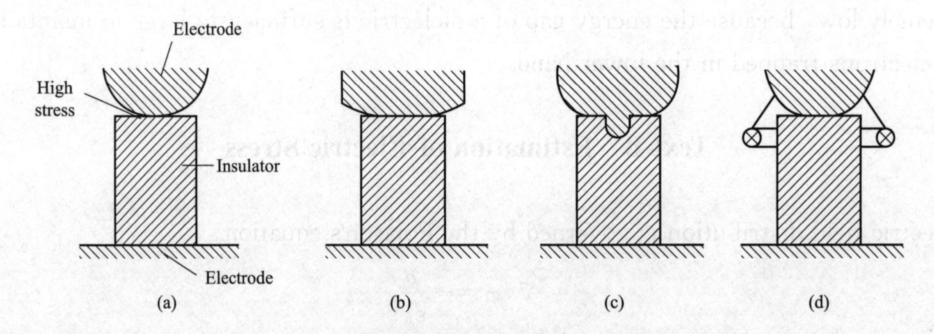

Fig. 5. 1 Control of stress at an electrode edge

(a) hemispherical electrode; (b) the electrode diameter is larger than the insulator diameter; (c) electrode protruding into insulator; (d) electrode with grading ring

Electric Field

A brief review of the concepts of electric fields is presented, since it is essential for high voltage engineers to have knowledge of the field intensities in various media under electric stresses. It also helps in choosing proper electrode configurations and economical dimensioning of the insulation, such that highly stressed regions are not formed and reliable operation of the equipment results in its anticipated life.

The field intensity E at any location in an electrostatic field is the ratio of the force on an infinitely small charge at that location to the charge itself as the charge decreases to zero. The force F on any charge q at that point in the field is given by

$$F = qE$$

The electric flux density D associated with the field intensity E is

$$D = \varepsilon E$$

where ε is the permittivity of the medium in which the electric field exists. The work done on a charge when moved in an electric field is defined as the potential. The potential φ is equal to

$$\varphi = -\int_l E \mathrm{d}l$$

where l is the path through which the charge is moved.

Several relationships between the various quantities in the electric field are summarized as follows:

$$\iint_S E \cdot \mathrm{d}S = \frac{q}{\varepsilon_0} \text{(Gauss theorem)}$$

$$\nabla \cdot D = \rho \text{(Charge density)}$$

where S is the closed surface containing charge q.

Words and Phases

space charge 空间电荷

permittivity *n.* 介电常数

operator *n.* 算子

Laplacian *n.* 拉普拉斯算子

scalar *n.* 标量

electrolytic tank 电解槽

equipotential *n.* 等势线，等电位线

curvilinear *adj.* 曲线的，弯曲的；*n.* 曲线

electric stress 电介质应力，静电应力

symmetry *n.* 对称

electric flux density 电通量密度

electrostatic *adj.* 静电的，静电学的

Notes

1. Once the voltage distribution of a given geometry is established, it is easy to redesign the electrodes to minimize the stresses so that the onset of corona is prevented.

一旦确定了给定结构上的电压分布，很容易重新设计电极以减少其静电应力，从而防止电晕放电的发生。

2. It also helps in choosing proper electrode configurations and economical dimensioning of the insulation, such that highly stressed regions are not formed and reliable operation of the equipment results in its anticipated life.

这也有助于正确选择电极结构和节省绝缘，以至于不能形成较高静电应力的区域，并使设备在预期寿命里可靠的运行。

Exercises

1. Answer the following questions according to the text

(1) What is the Laplace's equation when the space charge density $\rho=0$ in the region?

(2) Are there many methods available for determining the potential distribution? Say out the most commonly used methods.

(3) What important rules are there to sketch the field lines and equipotential?

(4) How to choose model to save the labor and time when redesigning the electrodes?

(5) Is the electric stress very much higher in the medium of lower permittivity when two dielectrics of widely different permittivities are in a series?

2. Translate the following sentences into Chinese according to the text

(1) In many practical cases, a good understanding of the problem is possible by using some simple rules to sketch the field lines and equipotential.

(2) This enhanced stress occurs at the electrode edges and one method of overcoming this is to increase the electrode diameter.

(3) A brief review of the concepts of electric fields is presented, since it is essential for high voltage engineers to have knowledge of the field intensities in various media under electric stresses.

(4) The field intensity E at any location in an electrostatic field is the ratio of the force on an infinitely small charge at that location to the charge itself as the charge decreases to zero.

(5) The work done on a charge when moved in an electric field is defined as the potential.

3. Translate the following paragraph into Chinese.

When several conductors are situated in an electric field with the conductors charged, a definite relationship exists among the potentials of the conductors, the charges on them, and the physical location of the conductors with respect to each other.

In a conductor, electrons can move freely under the influence of an electric field. This means that the charges are distributed inside the substance and over the surface such that, $E=0$ everywhere inside the conductor. Since $E=-\nabla\varphi=0$, it is necessary that φ is constant inside and on the surface of the conductor. Thus, the conductor is an equipotential surface.

Unit 2 Insulating Materials

Text A Dielectric Materials

1. Gas/Vacuum as Insulator

Air at atmospheric pressure is the most common gaseous insulation. The breakdown of air is of considerable practical importance to the design engineers of power transmission lines and power apparatus. Breakdown occurs in gases due to the process of collisional ionization. Electrons get multiplied in an exponential manner, and if the applied voltage is sufficiently large, breakdown occurs. In some gases, free electrons are removed by attachment to neutral gas molecules; the breakdown strength of such gases is substantially large. An example of such a gas with larger dielectric strength is sulphur hexafluoride (SF_6).

High pressure gas provides a reliable medium for high voltage insulation. Using gases at high pressures, field gradients up to 25 MV/m have been realized. Nitrogen (N_2) is the gas first used at high pressures because of its inertness and chemical stability, but its dielectric strength is the same as that of air. Other important practical insulating gases are carbon-dioxide (CO_2), dichlorodifluoromethane (CCl_2F_2) (popularly known as freon), and sulphur hexafluoride (SF_6). Ideally, vacuum is the best insulator with field strengths up to 10^7 V/cm, limited only by emissions from the electrode surfaces. Vacuum insulation is used in particle accelerators, X-ray and field emission tubes, electron microscopes, capacitors, and circuit breakers.

2. Liquid Breakdown

Liquids are used in high voltage equipment to serve the dual purpose of insulation and heat conduction. They have the advantage that a puncture path is self-healing. Temporary failures due to overvoltages arehealed quickly by liquid flow to the attacked area.

Under actual service conditions, the breakdown strength reduces considerably due to the presence of impurities. The breakdown mechanism in the case of very pure liquids is the same as the gas breakdown, but in commercial liquids, the breakdown mechanisms are significantly altered by the presence of the solid impurities and dissolved gases.

Petroleum oils are the commonest insulating liquids. However, fluorocarbons, silicones, and organic esters including castor oil are used in significant quantities. A number of considerations enter into the selection of any dielectric liquid. The important electrical properties of the liquid include the dielectric strength, conductivity, gas content, viscosity, permittivity, dissipation factor, stability, etc. Because of their low dissipation factor and other excellent characteristics, polybutenes are being increasingly used in the electrical industry. Silicones are particularly useful in transformers and capacitors and can be used at temperatures of 200 ℃ and higher. Castor oil is a good dielectric for high voltage energy storage capacitors because of its high corona resistance, high permittivity, and non-toxicity. In practical applications liquids are normally used at voltage stresses of about 50～60 kV/cm when the equipment is continuously operated. On the other hand, in applications like high voltage bushings, where the liquid only fills up the voids in the solid dielectric, it can be used at stresses as high as 100～200 kV/cm.

3. Solid Breakdown

If the solid insulating material is truly homogeneous and is free from imperfections, its breakdown stress will be as high as 10 MV/cm. However, in practice, the breakdown fields obtained are very much lower than this value. The breakdown occurs due to many mechanisms. In general, the breakdown occurs over the surface than in the solid itself, and the surface insulation failure is the most frequent cause of trouble in practice.

The breakdown of insulation can occur due to mechanical failure caused by the mechanical stresses produced by the electrical fields.

On the other hand, breakdown can also occur due to chemical degradation caused by the heat generated due to dielectric losses in the insulating material. This process is cumulative and is more severe in the presence of air and moisture.

When breakdown occurs on the surface of an insulator, it can be a simple flashover on the surface. Surface flashover results in the degradation of the material and normally occurs when the solid insulator is immersed in a liquid dielectric. Surface flashover, as already mentioned, is the most frequent cause of trouble in practice.

Porcelain insulators for use on transmission lines must therefore be designed to have a long path over the surface. Surface contamination of electrical insulation exists almost everywhere to some degree. In porcelain high voltage insulators of the suspension type, the length of the path over the surface will be 20 to 30 times greater than that through the solid. Even there, surface breakdown is the commonest form of failure.

So far, the various mechanisms that cause breakdown in dielectrics have been discussed. It is the intensity of the electric field that determines the onset of breakdown and the rate of increase of current before breakdown. Therefore, it is very essential that the electric stress should be properly estimated and its distribution known in a high voltage apparatus. Special care should be exercised in eliminating the stress in the regions where it is expected to be maximum, such as in the presence of sharp points.

4. Solid-liquid Insulating Systems

Impregnated-paper insulation constitutes one of the earliest insulating systems employed in electrical power apparatus and cables. Although in some applications alternate solid- or compressed-gas insulating systems are now being used, the impregnated-paper system still constitutes one of the most reliable insulating systems available. Impregnation of the paper results in a cavity-free insulating system, thereby eliminating the occurrence of partial discharges that inevitably lead to deterioration and breakdown of the insulating system. The cellulose structure of paper has a finite acidity content as well as a colloidal water, which is held by hydrogen bonds. Consequently, impregnated cellulose base papers are characterized by somewhat more elevated tanδ values in the order of 2×10^3 at 30 kV/cm. The liquid impregnants employed are either mineral oils or synthetic fluids. Since the permittivity of these fluids is normally about 2.2 and that of dried cellulose about 6.5～10, the resulting permittivity of the impregnated paper is approximately 3.1～3.5.

Lower-density cellulose papers have slightly lower dielectric losses, but the dielectric breakdown strength is also reduced. The converse is true for impregnated systems utilizing higher-density papers. The general chemical formula of cellulose paper is $C_{12}H_{20}O_{10}$. If the paper is heated beyond 200 ℃, the chemical structure of the paper breaks down even in the absence of external oxygen, since the latter is readily available from within the cellulose molecule. To avert this process from occurring, cellulose papers are ordinarily not used beyond 100 ℃.

In an attempt to reduce the dielectric losses in solid-liquid systems, cellulose papers have been substituted in some applications by synthetic papers (cf. Tab. 5.1). For example in extra-high-voltage cables, cellulose paper-polypropylene composite tapes have been employed. A partial paper content in the composite tapes is necessary both to retain some of the impregnation capability of a porous cellulose paper medium and to maintain the relative ease of cellulose tape sliding capability upon bending. In transformers the synthetic nylon or polyamide paper (nomex) has been used both in film and board form. It may be continuously operated at temperatures up to 220 ℃.

Tab. 5.1 Electrical properties of taped solid-liquid insulations

Tape	Impregnating Liquid	Average Voltage Stress (kV/cm)	tanδ at Room Temperature	tanδ at Operating Temperature
Kraft paper	Mineral oil	180	3.8×10^{-3} at 23 ℃	5.7×10^{-3} at 85 ℃
Kraft paper	Silicone liquid	180	2.7×10^{-3} at 23 ℃	3.1×10^{-3} at 85 ℃
Paper-polypropylene-paper (PPP)	Dodecyl benzene	180	9.8×10^{-4} at 18 ℃	9.9×10^{-4} at 100 ℃
Kraft paper	Polybutene	180	2.0×10^{-3} at 25 ℃	2.0×10^{-3} at 85 ℃

Words and Phases

collisional *adj*. 碰撞引起的

exponential *adj*. 指数的，幂数的

neutral gas molecules 惰性气体分子

sulphur hexafluoride 六氟化硫

inertness *n*. 惰性

carbon-dioxide 二氧化碳

dichlorodifluoromethane (freon) 二氯二氟甲烷（氟利昂）

impurity *n*. 杂质

fluorocarbon *n*. 碳氟化合物

silicone *n.* 硅树脂
castor oil 蓖麻油
organic ester 有机酯
conductivity *n.* 传导性，传导率
viscosity *n.* 黏性，黏度
dissipation factor 耗散因数，介质损耗角
polybutene *n.* 聚丁烯
non-toxicity *n.* 无毒性
bushing *n.* 套管
homogeneous *adj.* 均匀的，均质的
degradation *n.* 降解
flashover *n.* 闪络，跳火
porcelain *n.* 瓷器，瓷 adj. 瓷制的
contamination *n.* 污染，污染物
suspension *n.* 悬浮，悬挂
impregnated-paper 树脂浸渍纸
compressed-gas 压缩气体
impregnation *n.* 浸渍
cavity-free 有空腔的
cellulose *n.* 纤维素
acidity *n.* 酸度，酸性
colloidal *adj.* 胶状的，胶质的；*n.* 胶性，胶度
hydrogen bond 氢键
mineral oil 矿物油
polypropylene *n.* 聚丙烯
porous *n.* 多孔的，能渗透的
polyamide paper（nylon） 聚酰胺纸（尼龙）
kraft paper 牛皮纸
dodecyl benzene 十二烷基苯
nomex *n.* 高熔点芳香族聚酰胺

Notes

1. Impregnation of the paper results in a cavity-free insulating system, thereby eliminating the occurrence of partial discharges that inevitably lead to deterioration and breakdown of the insulating system.

浸渍纸形成一个空腔绝缘体系，从而消除部分电荷的产生，但却不可避免的导致了绝缘体系的劣化和击穿。

2. A partial paper content in the composite tapes is necessary both to retain some of the impregnation capability of a porous cellulose paper medium and to maintain the relative ease of cellulose tape sliding capability upon bending.

在合成绝缘带中必须有一定的纸含量，以保持多孔纤维纸介质一定的浸渍能力和纤维纸带弯曲时相对灵活的弹性力。

Exercises

1. Answer the following questions according to the text

(1) How does the breakdown in gases occur?

(2) What are some advantages and disadvantages to use nitrogen (N_2) at high pressures as insulation?

(3) What aspects is vacuum insulation used in?

(4) What are the important electrical properties of the liquid?

(5) What does surface flashover result in?

2. Translate the following sentences into Chinese according to the text

(1) The breakdown of air is of considerable practical importance to the design engineers of power transmission lines and power apparatus.

(2) Liquids are used in high voltage equipment to serve the dual purpose of insulation and heat conduction.

(3) If the solid insulating material is truly homogeneous and is free from imperfections, its breakdown stress will be as high as 10 MV/cm.

(4) It is the intensity of the electric field that determines the onset of breakdown and the rate of increase of current before breakdown.

(5) Lower-density cellulose papers have slightly lower dielectric losses, but the dielectric breakdown strength is also reduced.

3. Translate the following paragraph into Chinese

As the voltage is increased across a dielectric material, a point is ultimately reached beyond which the insulation will no longer be capable of sustaining any further rise in voltage and breakdown will ensue, causing a short to develop between the electrodes. If the dielectric consists of a gas or liquid medium, the breakdown will be self-healing in the sense that the gas or liquid will support anew a reapplication of voltage until another breakdown recurs. In a solid dielectric, however, the initial breakdown will result in a formation of a permanent conductive channel, which cannot support a reapplication of voltage. The dielectric breakdown processes are distinctly different for the three states of matter.

Text B Applications of Insulating Materials

There is no piece of electrical equipment that does not depend on electrical insulation in one form or other to maintain the flow of electric current in desired paths or circuits. If due to some reasons the current deviates from the desired path, the potential will drop. An example of this is a short circuit and this should always be avoided. This is done by proper choice and application of insulation wherever there is a potential difference between neighboring conducting bodies that carry current.

1. Applications in Power Transformers

Power transformers are the first to encounter lightning and other high voltage surges. The transformer insulation has to withstand impulse voltages many times higher than the power frequency operating voltages. The transformer insulation is broadly divided into

- conductor or turn-to-turn insulation
- coil-to-coil insulation
- low voltage coil-to-earth insulation
- high voltage coil-to-low voltage coil insulation
- high voltage coil-to-ground insulation

The low voltage coil-to-ground and the high voltage coil-to-low voltage coil insulations normally consist of solid tubes combined with liquid or gas filled spaces. The liquid or gas in the spaces help to remove the heat from the core and coil structure and also help to improve the insulation strengths. In the large transformers paper or glass tape is wrapped on the rectangular conductors. In the case of layer to layer, coil-to-coil and coil-to-ground insulations, Kraft paper is used in smaller transformers, whereas thick radial spacers made of pressboard, glass fabric, or porcelain are used in the case of higher rating transformers.

Of all the materials, oil impregnated paper, and pressboard are extensively used in liquid filled transformers. The lack of thermal stability at higher temperatures limits the use of this type of insulation to be used continuously up to 105 ℃. Paper and its products absorb moisture very rapidly from the atmosphere, and hence this type of insulation should be kept free of moisture during its life in a transformer.

Transformer oil provides the required dielectric strength and insulation and also cools the transformer by circulating itself through the core and the coil structure. The transformer oil, therefore, should be in the liquid state over the complete operating range of temperatures between −40 ℃ and +50 ℃. The oil gets oxidized when exposed to oxygen at high temperatures, and the oxidation results in the formation of peroxides, water, organic acids and sludge. These products cause chemical deterioration of the paper insulation and the metal parts of the transformer. Sludge being heavy, reduces the heat transfer capabilities of the oil, and also forms as a heat insulating layer on the coil structure, the core and the tank walls. In present-day transformers the effects of oxidation are minimized by designing them such that access to oxygen itself is limited. This is done by the use of sealed transformers, by filling the air space with nitrogen gas, and providing oxygen absorbers like activated clay or alumina.

When an arc discharge occurs inside a transformer, the oil decomposition occurs. The decomposition products consist of hydrogen and gaseous hydrocarbons which may lead to explosion. And hence, oil insulated transformers are seldom used inside buildings or other hazardous locations like mines. Under such conditions dry type and askarel or sulphur hexafluoride (SF_6) gas filled transformers are used. Askarel is a fireproof liquid and is the generic name for a number of synthetic chlorinated aromatic hydrocarbons. These are more stable to oxidation and do not form acids or sludge. Under arcing they are very stable and do not give rise to inflammable gases. However they give out hydrochloric acid which is toxic and which attacks the paper insulation. This is removed by using tin or tetraphenyl. However, if the arc is very heavy, the hydrochloric acid cannot be absorbed completely. For these reasons SF_6 gas insulated transformers are popular. Also, askarel cannot be used in high voltage transformers, because the impulse strength of askarel impregnanted paper is very low compared to that of oil impregnated paper. Moreover, its dielectric strength deteriorates rapidly at high voltages and at high frequencies liberating hydrochloric acid.

Even today there is no perfect all purpose transformer fluid. In recent years, progress has been made with the use of fluorocarbon liquids and SF_6 gas. However, these liquids have not become very popular because of their high cost.

2. Applications in Circuit Breakers

A circuit breaker is a switch which automatically interrupts the circuit when a critical current or voltage rating is exceeded. AC currents are considerably easier to interrupt than DC currents. AC current interruption generally requires first to substitute an arc for part of the metallic circuit and then its deionization when the current goes through zero, so that the arc will not reestablish again.

Circuit breakers are also divided into two categories, namely, the low voltage and high voltage types. Low voltage breakers use synthetic resin mouldings to carry the metallic parts. For higher temperatures ceramic parts are used. When the arc is likely to come into contact with moulded parts, some special kind of alkyd resins are used because of their greater arc resistance. The high voltage circuit breakers are further classified into air circuit breakers and oil circuit breakers. Many insulating fluids are suitable for arc extinction and the choice of the fluid depends on the rating and type of the circuit breaker. The insulating fluids commonly used are atmospheric air, compressed air, high vacuum, SF_6 and oil. In some ancillary equipment used with circuit breakers, the fluid serves the purpose of providing only insulation. Many insulants are available for this purpose.

The oils used in circuit breakers normally have the same characteristics as transformer oil. In circuit breakers oil serves an additional purpose of interrupting the arc. Since the gases (mainly hydrogen) help to extinguish the arc, a liquid which generates the maximum amount of the gas for one unit of arc energy is preferred. Transformer oil possesses these characteristics. Many other oils have been tried but with no success. Askarels produce large quantities of toxic and corrosive products.

The circuit breaker bushings of lower voltage ratings may consist of solid cylinders of porcelain and shellac wrapped on the current carrying electrode. High voltage bushings of voltages of 66 kV and above are filled with oil. The constructional details vary widely. In certain designs, the system of coaxial porcelain is used with space between them filled with oil.

Words and Phases

spacer *n.* 垫片，垫板
pressboard *n.* 纸板
oxidation *n.* 氧化
peroxide *n.* 过氧化物
sludge *n.* 油泥，（油罐底）酸渣
sealed *n.* 封闭的，密封的
activated clay 活性黏土
alumina *n.* 氧化铝
decomposition *n.* 分解
hydrocarbon *n.* 烃类，碳氢化合物
askarel 爱斯开勒电解液体，合成绝缘液（当电火花 使这种液体分解时，它会放出不燃气体）
fireproof *adj.* 耐火的，防火的
chlorinated aromatic hydrocarbon 氯代芳烃类
inflammable *adj.* 易燃的
toxic *adj.* 有毒的

tetraphenyl *n.* 四苨基
metallic *adj.* 金属（性）的
deionization *n.* 除去离子，消除电离
resin *n.* 树脂
moulding *n.* 压制
come into contact with *v.* 跟……接触，同……接触起来
moulded *adj.* 成型的，模压的
arc extinction 灭弧，消弧
insulant *n.* 绝缘体，绝缘材料
ancillary *adj.* 补助的，副的
corrosive *adj.* 腐蚀的，腐蚀性的；*n.* 腐蚀物，腐蚀剂
cylinder *n.* 柱体，汽缸
alkyd resin 醇酸树脂，聚酯树脂
shellac *n.* 虫胶，紫胶
coaxial *adj.* 同轴的，共轴的

Exercises

1. Answer the following questions according to the text

(1) What categories does the transformer insulation consist of?

(2) What does the oil oxidation produce when exposed to oxygen at high temperatures?

(3) Why are oil insulated transformers seldomly used inside buildings and other hazardous locations like mines?

(4) Where are the circuit breakers used?

(5) How to interrupt AC current?

2. Translate the following sentences into Chinese according to the text

(1) The liquid or gas in the spaces help to remove the heat from the core and coil structure and also help to improve the insulation strengths.

(2) The lack of thermal stability at higher temperatures limits the use of this type of insulation to be used continuously up to 105 ℃.

(3) Even today there is no perfect all-purpose transformer fluid.

(4) Circuit breakers are also divided into two categories, namely, the low voltage and high voltage types.

(5) High voltage bushings of voltages of 66 kV and above are filled with oil.

3. Translate the following paragraph into Chinese

As already mentioned, the latest trend in capacitor manufacture is to replace paper by polypropylene plastic films. This results in a drastic reduction in size. Its use results in cheaper capacitors for high voltage ratings because of its high working stress. As regards impregnants, askarels are harmful to the environment and hence are being banned. The latest trend is to develop other types of materials. With this in view, research is being directed towards the use of oils like castor oil.

Unit 3 Insulation Testing of Electrical Apparatus

Text A Measurement of High Voltages

In industrial testing and research laboratories, it is essential to measure the voltages accurately, ensuring perfect safety to the personnel and equipment. Hence a person handling the equipment as well as the metering devices must be protected against overvoltages and also against any induced voltages due to stray coupling. Electromagnetic interference is a serious problem in impulse voltage measurements, and it has to be avoided or minimized. Therefore, even though the principles of measurements may be same, the devices and instruments for measurement of high voltages differ vastly from the low voltage devices. Different devices used for high voltage measurements may be classified as in Tab. 5. 2.

Tab. 5. 2 High voltage measurement techniques

Type of voltage	Method or technique
DC voltages	Series resistance microammeter
	Resistance potential divider
	Generating voltmeters
power frequency voltages	Series impedance ammeters
	Potential dividers (resistance or capacitance type)
	Potential transformers
	Electrostatic voltmeters
AC high frequency voltages, impulse voltages, and other rapidly changing voltages	Potential dividers with a cathode ray oscillograph (resistive or capacitive dividers)
	Peak voltmeters

1. Measurement of High DC Voltages

Measurement of high DC voltages as in low voltage measurements, is generally accomplished by extension of meter range with a large series resistance. The net current in the meter is usually limited to one to ten microamperes for full-scale deflection. For very high voltages (1000 kV or more) problems arise due to large power dissipation, leakage currents and limitation of voltage stress per unit length, change in resistance due to temperature variations, etc. Hence, a resistance potential divider with an electrostatic voltmeter is sometimes better when high precision is needed. But potential dividers also suffer from the disadvantages stated above. Both series resistance meters and potential dividers cause current drain from the source. Generating voltmeters are high impedance devices and do not load the source. They provide complete isolation from the source voltage (high voltage) as they are not directly connected to the high voltage terminal and hence are safer.

2. Measurement of High AC Voltages

Measurement of high AC voltages employs conventional methods like series impedance voltmeters, potential dividers, potential transformers, or electrostatic voltmeters. But their designs are different from those of low voltage meters, as the insulation design and source loading are the important criteria.

3. Series Impedance Voltmeters

For power frequency AC measurements the series impedance may be a pure resistance or a reactance. Since resistances involve power losses, often a capacitor is preferred as a series reactance. Moreover, for high resistances, the variation of resistance with temperature is a problem, and the residual inductance of the resistance gives rise to an impedance different from its ohmic resistance. High resistance units for high voltages have stray capacitances and hence a unit resistance will have an equivalent circuit as shown in Fig. 5. 2. At any frequency ω of the AC voltage, the impedance of the resistance R is

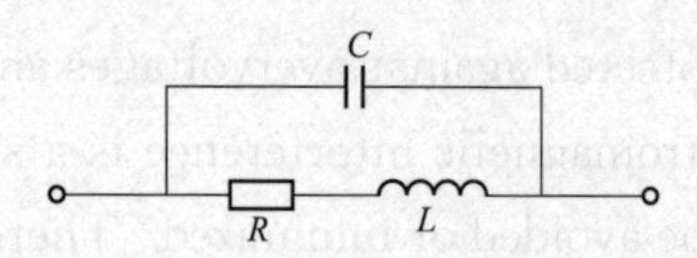

Fig. 5. 2 Simplified lumped parameter equivalent circuit of a high ohmic resistance R

L—Residual inductance;

C—Residual capacitance

$$Z=\frac{R+\mathrm{j}\omega L}{(1-\omega^2 LC)+\mathrm{j}\omega CR}$$

If ωL and ωC are small compared to R,

$$Z\approx R\left[1+\mathrm{j}\left(\frac{\omega L}{R}-\omega CR\right)\right]$$

and the total phase angle is

$$\varphi\approx\left(\frac{\omega L}{R}-\omega CR\right)$$

This can be made zero and independent of frequency, if

$$L/C=R^2$$

For extended and large dimensioned resistors, this equivalent circuit is not valid and each elemental resistor has to be approximated with this equivalent circuit. The entire resistor unit then has to be taken as a transmission line equivalent, for calculating the effective resistance. Also, the ground or stray capacitance of each element influences the current flowing in the unit, and the indication of the meter results in an error. The equivalent circuit of a high voltage resistor neglecting inductance and the circuit of compensated series resistor using guard and tuning resistors is shown in Fig. 5. 3 (a) and (b) respectively. Stray capacitance to ground effects [refer Fig. 5. 3 (b)] can be removed by shielding the resistor R by guard resistor R_s, which shunts the actual resistor but does not contribute to the current through the instrument. By tuning the resistors R_a, the shielding resistor end potentials may be adjusted so that the resulting compensation currents between the shield and the measuring resistors provide a minimum phase angle.

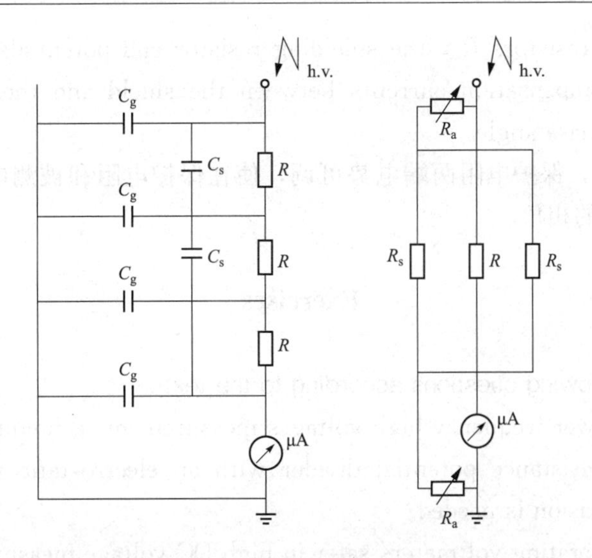

Fig. 5.3 Extended series resistance for high AC voltage measurements

(a) Extended series resistance with inductance neglected;

(b) Series resistance with guard and tuning resistances

C_g—Stray capacitance to ground; C_s—Winding capacitance; R—Series resistor;

R_s—Guard resistor; R_a—Tuning resistor

Words and Phases

metering device 测量装置，计量仪表

induced *adj*. 感应的，感生的

stray coupling 杂散耦合，寄生耦合

micro-ammeter 微安计，微安表

potential divider 分压器

series impedance 串联阻抗

ammeter *n*. 电流表，安培计

cathode ray oscillograph 阴极射线示波器

net current 净电流

microampere *n*. （电流单位）微安培

full-scale deflection 满刻度偏转，最大偏转

power dissipation 功率耗散，功率消耗

precision *n*. 精确，精密度，精度

source *n*. 源，电源

power loss 功率损耗

residual *adj*. 剩余的，残留的

ohmic *adj*. 电阻性的，欧姆性的

lumped parameter 集中参数测量，集总参量

shunt *v*. 分流，*n*. 分路器，分流器

Notes

1. Stray capacitance to ground effects [refer Fig. 3.2 (b)] can be removed by shielding the resistor R by guard resistor R_s, which shunts the actual resistor but does not contribute to the current through the instrument.

如图 3.2（b）所示，对地的寄生电容效应通过保护电阻 R_s 屏蔽电阻 R 而消除。R_s 只分流实际电阻而不通过仪表贡献电流。

2. By tuning the resistors R_a, the shielding resistor end potentials may be adjusted so that the resulting compensation currents between the shield and the measuring resistors provide a minimum phase angle.

通过调节电阻 R_a，保护电阻两端电势可调，使在保护电阻和被测电阻间相应变化的补偿电流获得一个最小的相角。

Exercises

1. Answer the following questions according to the text

(1) What are power frequency high voltages measurement techniques?

(2) Why is a resistance potential divider with an electrostatic voltmeter sometimes better when high precision is needed?

(3) Why are generating voltmeters safer in high DC voltage measurements?

(4) What conventional methods does measurement of high AC voltages employ?

(5) What is the impedance of the resistance R, at any frequency ω of the AC voltage?

2. Translate the following sentences into Chinese according to the text

(1) In industrial testing and research laboratories, it is essential to measure the voltages accurately, ensuring perfect safety to the personnel and equipment.

(2) Therefore, even though the principles of measurements may be same, the devices and instruments for measurement of high voltages differ vastly from the low voltage devices.

(3) The net current in the meter is usually limited to one to ten microamperes for full-scale deflection.

(4) But their designs are different from those of low voltage meters, as the insulation design and source loading are the important criteria.

(5) The entire resistor unit then has to be taken as a transmission line equivalent, for calculating the effective resistance.

3. Translate the following paragraph into Chinese

A fault in an electrical power system is the unintentional and undesirable creation of a conducting path (a short circuit) or a blockage of current (an open circuit). The short circuit fault is typically the most common and is usually implied when most people use the term fault. We restrict our comments to the short circuit fault.

The causes of faults include lightning, wind damage, trees falling across lines, vehicles colliding with towers or poles, birds shorting out lines, aircraft colliding with lines, vandalism, small animals entering switchgear, and line breaks due to excessive ice loading. Power system faults may be categorized as one of four types: single line-to-ground, line-to-line, double line-to-ground and balanced three-phase. The first three types constitute severe unbalanced operating conditions.

Text B Non-Destructive Testing of Materials and Electrical Apparatus

Electrical insulating materials are used in various forms to provide insulation for the apparatus. The insulating materials may be solid, liquid, gas, or even a combination of these such as paper impregnated with oil. These materials should possess good insulating properties over a wide range of operating parameters, such as a wide temperature range and a wide frequency range. Since it is difficult to test the quality of an insulating material after it forms part of equipment, suitable tests must be done to ensure their quality in the said ranges of operation. Also, these tests are devised to ensure that the material is not destroyed as in the case of high voltage testing.

1. Measurement of DC Resistivity

1.1 Specimens and Electrodes

The specimen shape and the electrode arrangement should be such that the resistivity can be easily calculated. For a solid specimen, the preferable shape is a flat plate with plane and parallel surfaces, usually circular. The specimens are normally in the form of discs of 5 to 10 cm diameter and 3 to 12 mm thickness.

If the electrodes are arranged to be in contact with the surfaces of the specimen, the measured resistance will be usually greater due to the surface conductivity effects. Often, a three electrode arrangement shown in Fig. 5.4 is used. The electrode which completely covers the surface of the specimen is called the "unguarded electrode" and is connected to the high voltage terminal. The third electrode which surrounds the other measuring electrode is connected to a suitable terminal of the measuring circuit. The width of this "guard electrode" must be at least twice the thickness of the specimen, and the unguarded electrode must extend to the outer edge of the guard electrode. The gap between the guarded and guard electrodes should be as small as possible. The effective diameter of the guarded electrode is greater than the actual diameter and is given as follows.

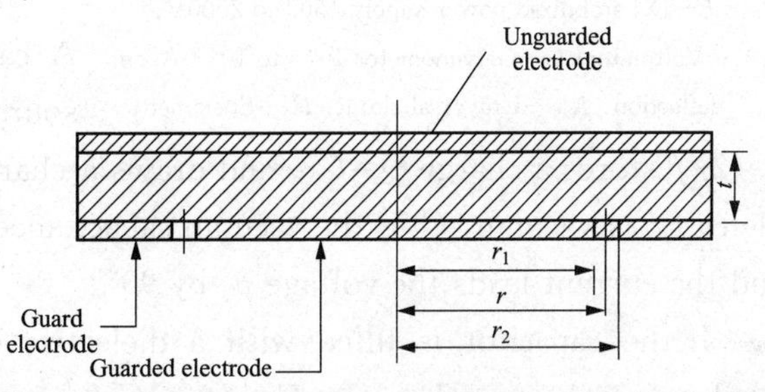

Fig. 5.4 The three electrode system

Let r_1, r_2, and r be the radius of the guarded electrode, guard electrode including the gap, and the effective radius of the guarded electrode. Let the gap width is g and the specimen thickness is t.

$$r = r_1 + \frac{g}{2} - \delta$$

$$\frac{\delta}{t} = \frac{2}{\pi} \ln \cosh\left(\frac{\pi g}{4t}\right)$$

1.2 Measuring Circuits

A simple measuring circuit for the measurement of resistance is shown in Fig. 5.5. The galvanometer is first calibrated by using a standard resistance of 1 to 10 MΩ (±0.5% or ±1%). If necessary, a standard universal shunt is used with the galvanometer. The deflection in cm per microampere of current is noted. The specimen (R_p) is inserted in the current as shown in Fig. 5.5, and maintaining the same supply voltage, the galvanometer current is observed by adjusting the universal shunt, if necessary. The galvanometer gives a maximum sensitivity of 10^{-9} A/cm deflection and a DC amplifier has to be used along with the galvanometer for higher sensitivities (up to 10^{-12} to 10^{-13} A/cm).

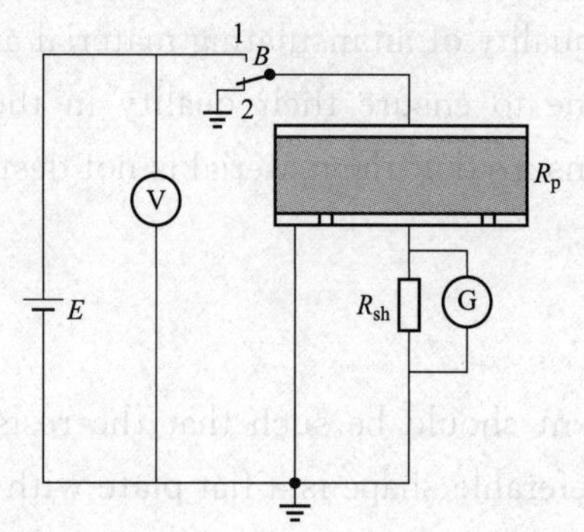

Fig. 5.5 DC galvanometer arrangement

E—DC stabilized power supply, 500 to 2000V; V—Voltmeter; G—Galvanometer 10^{-8} to 10^{-9} A/cm deflection; R_{sh}—Universal shunt; R_p—Specimen

The resistance of the specimen is given by

$$R_p = \frac{U}{(D \times G)}$$

where D is deflection in cm (with specimen), and G is galvanometer sensitivity.

2. Measurement of Dielectric Loss Factor

A capacitor connected to a sinusoidal voltage source$u = u_0 e^{j\omega t}$ with an angular frequency $\omega = 2\pi f$ stores a charge $Q = C_0 u$ and draws a charging current $I_c = dQ/dt = j\omega C_0 u$. When the dielectric is vacuum, C_0 is the vacuum capacitance or geometric capacitance of the condenser, and the current leads the voltage v_C by 90°.

If the capacitor is filled with a dielectric of permittivity ε', the capacitance of the condenser is increased to $C = C_0\varepsilon'/\varepsilon_0 = C_0 K'$, where K' is the relative permittivity of the material with respect to vacuum.

Under these conditions, if the same voltage u is applied, there will be a charging current I_c and loss component of the current I_l. I_l will be equal to G_u where G represents the conductance of the dielectric material. The total current $I = I_c + I_l = (j\omega C + G)\ U$. The current leads the voltage by an angle θ which is less than 90°. The loss angle δ is equal to $(90° - \theta)$. The phasor diagrams of an ideal capacitor and a capacitor with a lossy dielectric are shown in Figs. 5.6 (a) and (b).

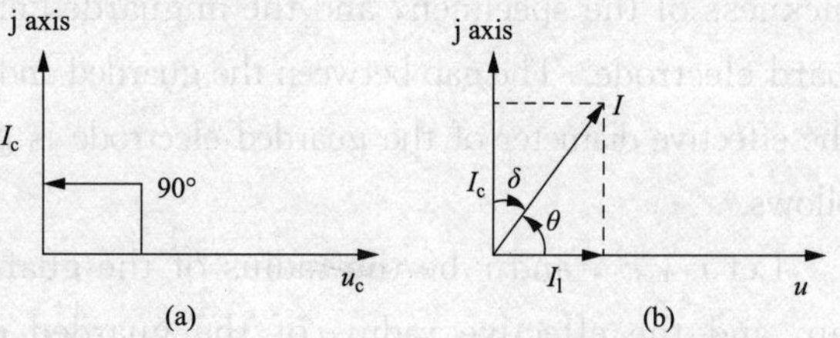

Fig. 5.6 Capacitor phasor diagrams

(a) Ideal capacitor; (b) Capacitor with a lossy dielectric

It would be premature to conclude that the dielectric material corresponds to an R-C parallel circuit in electrical behaviour. The frequency response of this circuit which can be expressed as the ratio of the loss current to the charging current, i. e. the loss tangent is

$$\tan\delta = D = \frac{I_1}{I_c} = \frac{1}{\omega CR}$$

3. Power Frequency Measurement Methods — High Voltage Schering Bridge

In the power frequency range (25 to 100 Hz) schering bridge is a very versatile and sensitive bridge and is readily suitable for high voltage measurements. The stress dependence of K' or ε_r and tan δ can be readily obtained with this bridge.

The schematic diagram of the bridge is shown in Fig. 5. 7. The lossy capacitor or capacitor with the dielectric between electrodes is represented as an imperfect capacitor of capacitance C_x together with a resistance r_x. The standard capacitor is shown as C_s which will usually have a capacitance of 50 to 500 μF. The variable arms are R_4 and C_3R_3. Balance is obtained when

$$\frac{Z_1}{Z_2}=\frac{Z_4}{Z_3}$$

where, $Z_1=r_x+\frac{1}{j\omega C_x}$, $Z_2=\frac{1}{j\omega C_s}$, $Z_3=\frac{R_3}{1+j\omega C_3R_3}$, and $Z_4=R_4$.

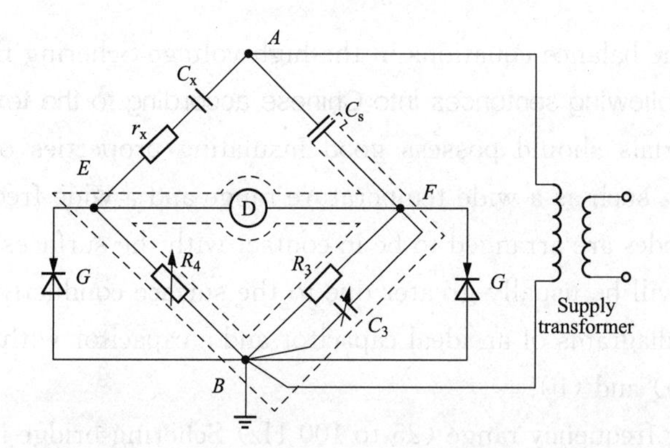

Fig. 5. 7 Schematic diagram of a Schering bridge

- - -dotted line is the shielding arrangement. Shield is connected to B, (the ground)

The balance equations are

$$C_x=\frac{R_3}{R_4}C_s\text{; and } r_x=\frac{C_3}{C_2}R_1$$

The loss angle is

$$\tan\delta_x = \omega C_x R_x = \omega C_3 R_3$$

Words and Phases

resistivity *n.* 电阻率，电阻系数

preferable *adj.* 优越的，更可取的，较好的

discs *n.* 圆板，圆盘

guard electrode 保护电极

guarded electrode 屏蔽电极
galvanometer *n.* 电流计，检流计
calibrate *v.* 校准
sensitivity *n.* 灵敏度，敏感性
angular *adj.* 有角的
geometric *adj.* 几何的，几何学的
condenser *n.* 电容器
lossy *adj.* 有损的；*n.* 有损耗的
premature *adj.* 过早的
loss tangent 损耗角正切，损耗因数
Schering bridge 西林电桥
schematic *adj.* 图解的，示意的

Exercises

1. Answer the following questions according to the text

(1) What is the preferable shape for a solid specimen?

(2) Which part is called the "unguarded electrode" in the three electrode system?

(3) How to calibrate the galvanometer in the measurement of resistance?

(4) What is the capacitance of the condenser if the capacitor is filled with a dielectric of permittivity ε'?

(5) Which are the balance equations in the high voltage Schering Bridge?

2. Translate the following sentences into Chinese according to the text

(1) These materials should possess good insulating properties over a wide range of operating parameters, such as a wide temperature range and a wide frequency range.

(2) If the electrodes are arranged to be in contact with the surfaces of the specimen, the measured resistance will be usually greater due to the surface conductivity effects.

(3) The phasor diagrams of an ideal capacitor and a capacitor with a lossy dielectric are shown in Figs. 5.6 (a) and (b).

(4) In the power frequency range (25 to 100 Hz) Schering bridge is a very versatile and sensitive bridge and is readily suitable for high voltage measurements.

(5) The standard capacitor is shown as C_s which will usually have a capacitance of 50 to 500 μF.

3. Translate the following paragraph into Chinese

Earlier the testing of insulators and other equipment was based on the insulation resistance measurements, dissipation factor measurements and breakdown tests. It was observed that the dissipation factor ($\tan\delta$) was voltage dependent and hence became a criterion for the monitoring of the high voltage insulation. In further investigations it was found that weak points in an insulation like voids, cracks, and other imperfections lead to internal or intermittent discharges in the insulation. These imperfections being small were not revealed in capacitance measurements but were revealed as power loss components in contributing for an increase in the dissipation factor. In modern terminology these are designated as "partial discharges" which in course of time reduce the strength of insulation leading to a total or partial failure or breakdown of the insulation.

Unit 4 Testing, Planning and Layout of High Voltage Laboratories

Text A High Voltage Testing of Electrical Apparatus

It is essential to ensure that the electrical equipment is capable of withstanding the over voltages that are met with in service. The over voltages may be either due to natural causes like lightning or system originated ones such as switching or power frequency transient voltages. Hence, testing for overvoltages is necessary.

1. Testing of Insulators and Bushings

1.1 Tests on Insulators

• Power Frequency Tests

(1) Dry and wet flashover tests

In these tests theAC voltage of power frequency is applied across the insulator and increased at a uniform rate of about 2 per cent per second of 75% of the estimated test voltage, to such a value that a breakdown occurs along the surface of the insulator. If the test is conducted under normal conditions without any rain or precipitation, it is called "dry flashover test". If the test is done under conditions of rain, it is called "wet flashover test". In general, wet tests are not intended to reproduce the actual operating conditions, but only to provide a criterion based on experience that a satisfactory service operation will be obtained. The test object is subjected to a spray of water of given conductivity by means of nozzles. The spray is arranged such that the water drops fall approximately at an inclination of 45° to the vertical. The test object is sprayed for at least one minute before the voltage application, and the spray is continued during the voltage application.

(2) Wet and dry withstand tests (one minute)

In these tests, the voltage specified in the relevant specification is applied under dry or wet conditions for a period of one minute with an insulator mounted as in service conditions. The test piece should withstand the specified voltage.

• Impulse Tests

(1) Impulse withstand voltage test

This test is done by applying standard impulse voltage of specified value under dry conditions with both positive and negative polarities of the wave. If five consecutive waves do not cause a flashover or puncture, the insulator is deemed to have passed the test. If two applications cause flashover, the object is deemed to have failed. If there is only one failure, additional ten applications of the voltage wave are made. If the test object has withstood the subsequent applications, it is said to have passed the test.

(2) Impulse flashover test

The test is done as above with the specified voltage. Usually, the probability of failure is determined for 40% and 60% failure values or 20% and 80% failure values, since it is difficult to adjust the test voltage for the exact 50% flashover values. The average value of the upper and the lower limits is taken. The insulator surface should not be damaged by these tests, but slight marking on its surface is allowed.

1.2　Testing of Bushings

• Power Frequency Tests

(1) Power factor—voltage test

In this test, the bushing is set up as in service or immersed in oil. It is connected such that the line conductor goes to the high voltage side and the tank goes to the detector side of the high voltage Schering bridge. Voltage is applied up to the line value in increasing steps and then reduced. The capacitance and power factor (or $\tan\delta$) are recorded at each step. The characteristic of power factor or $\tan\delta$ versus applied voltage is drawn. This is a normal routine test but sometimes may be conducted on percentage basis.

(2) Internal or partial discharge test

This test is intended to find the deterioration or failure due to internal discharges caused in the composite insulation of the bushing. This is done by using internal or partial discharge arrangement. The voltage versus discharge magnitude as well as the quadratic rate gives an excellent record of the performance of the bushing in service. This is now a routine test for high voltage bushings.

• Impulse Voltage Tests

(1) Full wave withstand test

The bushing is tested for either polarity voltages as per the specifications. Five consecutive full waves of standard waveform are applied, and, if two of them cause flashover, the bushing is said to have failed in the test. If only one flashover occurs, ten additional applications are done. The bushing is considered to have passed the test if no flashover occurs in subsequent applications.

(2) Chopped wave withstand and switching surge tests

The chopped wave test is sometimes done for high voltage bushings (220 kV and 400 kV and above). Switching surge flashover test of specified value is now-a-days included for high voltage bushings. The tests are carried out similar to full wave withstand tests.

2. Testing of Cables

Cables are very important electrical apparatus for transmission of electrical energy by underground means. They are also very important means for transmitting voltage signals at high voltages. For power engineers, large power transmission cables are of importance, and hence testing of power cables only is considered here.

2.1　Preparation of the Cable Samples

For overvoltage and withstand tests, samples have to be carefully prepared and terminated; otherwise, excessive leakage or end flashovers may occur during testing. The normal length of the cable sample used varies from about 50 cm to 10 m. The terminations are usually made by shielding the end conductor with stress shields. A few terminations are shown in Fig. 5. 8. During power factor tests, the cable ends are provided with shields so that the surface leakage current is avoided from the measuring circuits.

2. 2 High Voltage Tests on Cables

Cables are tested for withstand voltages using the power frequency AC, DC, and impulse voltages. At the time of manufacture, the entire cable is passed through a high voltage test at the rated voltage to check the continuity of the cable. As a routine test, the cable is tested applying an AC voltage of 2. 5 times the rated value for 10min. No damage to the cable insulation should occur. Type tests are done on cable samples using both high voltage DC and impulse voltages. The DC test consists of applying 1. 8 times the rated DC voltage of negative polarity for 30min, and the cable system is said to be fit, if it withstands the test. For impulse tests, impulse voltage of the prescribed magnitude as per specifications is applied, and the cable has to withstand five applications without any damage. Usually, after the impulse test, the power frequency dielectric power factor test is done to ensure that no failure occurred during the impulse test.

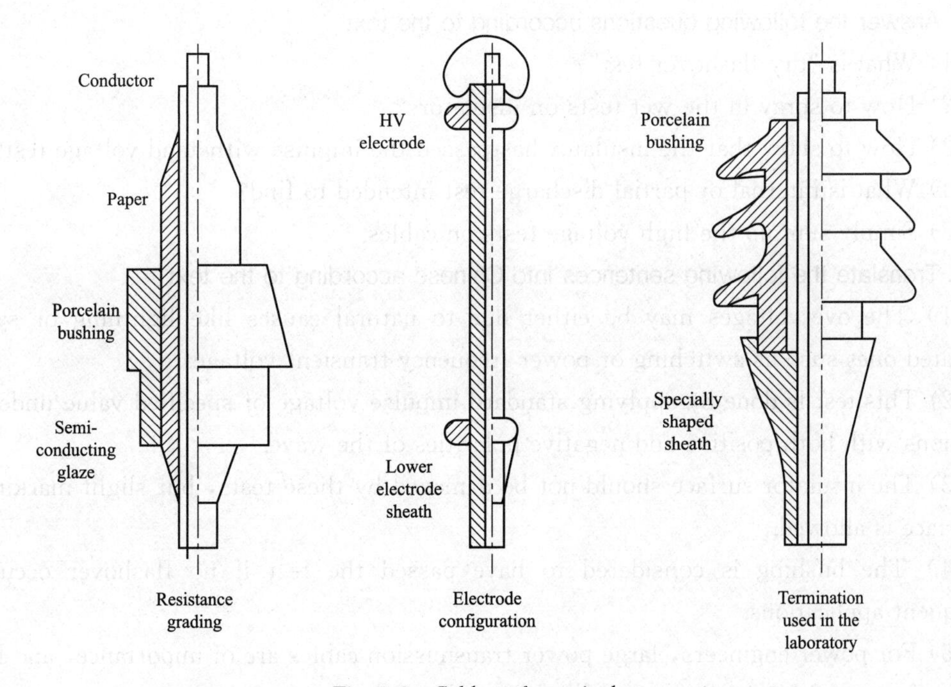

Fig. 5. 8 Cable and terminals

Words and Phases

nozzle *n.* 管口，喷嘴

inclination *n.* 倾度，倾角

mounted *adj*. 安装好的
piece *n*. 样本，样品，例子
consecutive *adj*. 连续的，连贯的
detector *n*. 探测器，检波器，传感元件
versus *prep*. 与…的比值，…与…的关系曲线
routine test 常规试验，出厂试验，例行试验
quadratic *adj*. 二次的；*n*. 二次方程式
as per 按照，根据
chopped wave 斩波，断续波
continuity *n*. 连续性，连贯性
type test 例行试验，型式试验

Notes

In these tests the AC voltage of power frequency is applied across the insulator and increased at a uniform rate of about 2 per cent per second of 75% of the estimated test voltage, to such a value that a breakdown occurs along the surface of the insulator.

在这个试验中，绝缘体两端加交流工频电压，在估计试验电压75%的基础上，以每秒2%的平均速度增加到使得沿绝缘体表面发生击穿的电压为止。

Exercises

1. Answer the following questions according to the text

(1) What is "dry flashover test"?

(2) How to spray in the wet tests on insulators?

(3) How to show that the insulator has passed the impulse withstand voltage test?

(4) What is internal or partial discharge test intended to find?

(5) Simply sum up the high voltage tests on cables.

2. Translate the following sentences into Chinese according to the text

(1) The overvoltages may be either due to natural causes like lightning or system originated ones such as switching or power frequency transient voltages.

(2) This test is done by applying standard impulse voltage of specified value under dry conditions with both positive and negative polarities of the wave.

(3) The insulator surface should not be damaged by these tests, but slight marking on its surface is allowed.

(4) The bushing is considered to have passed the test if no flashover occurs in subsequent applications.

(5) For power engineers, large power transmission cables are of importance, and hence testing of power cables only is considered here.

3. Translate the following paragraph into Chinese

Many electrical apparatuses like transformers, line conductors, rotating machines, etc. produce unwanted electrical signals in the radio and high frequency ranges. These signals arise due to corona discharges in air, internal or partial discharges in the insulation, sparking at commutators and brush gear in rotating machines, etc. It is important to see that the noise voltages generated in the radio and other transmission bands are limited to acceptable levels, and hence the radio interference voltage measurements are of importance. It has been found that the surface conditions of the overhead conductors subjected to high voltage stresses and varying atmospheric conditions greatly influence the magnitude of the noise voltage produced.

Text B Planning and Layout of High Voltage Laboratories

Industrial and economic development in the present world demands the use of more and more electrical energy which has to be transported over long distances in large quantities. Transportation of large amounts of power needs extra high voltage transmission lines. The AC transmission lines of ratings of 1000 kV or more have come into operation. Extensive studies are being made in different countries on the possible use of complex extra high voltage DC systems of ±500 kV and above.

This very fast development of power systems should be followed by system studies on equipment and service conditions which they have to fulfill. These conditions will also determine the values for test voltages of AC power frequency, impulse, or DC, under specific conditions.

High voltage laboratories are an essential requirement for making acceptance tests for the equipment that go into operation in the extra high voltage transmission systems. In addition, they are also used in the development work on equipment for conducting research, and for planning to ensure economical and reliable extra high voltage transmission systems. Here a brief review of the planning and layout of testing laboratories and some problems and limitations of the test techniques are presented.

1. Test Facilities Provided in High Voltage Laboratories

A high voltage laboratory is expected to carry out withstand and/or flashover tests at high voltages on the following transmission system equipment:

- Transformers
- Lightning arresters
- Isolators and circuit breakers
- Different types of insulators
- Cables
- Capacitors
- Line hardware and accessories
- Other equipment like reactors, etc

Different tests conducted on the above equipment are:

- Power frequency withstand tests-wet and dry
- Impulse tests
- DC withstand tests

➢ Switching surge tests

➢ Tests under polluted atmospheric conditions

➢ Partial discharge measurements

Apart from the above facilities which are needed for routine testing, the laboratories are expected to have facilities for studying dielectric properties of insulation and insulating materials.

2. Activities and Studies in High Voltage Laboratories

High voltage laboratories, in addition to conducting tests on equipment, are used for research and development works on the equipment. This includes determination of the safety factor for dielectrics and reliability studies under different atmospheric conditions such as rain, fog, industrial pollution, etc. , at voltage higher than the test voltage required. Sometimes, it is required to study problems associated with test lines and other equipment under natural atmospheric or pollution conditions, which cannot be done indoors.

Research activities usually include the following:

➢ Breakdown phenomenon in insulating media such as gases, liquids, solids, or composite systems.

➢ Withstand voltage on long gaps, surface flashover studies on equipment with special reference to the equipment and materials used in power systems.

➢ Electrical interference studies due to discharges from equipment operating at high voltages.

➢ Studies on insulationof HV, power systems.

➢ High current phenomenon such as electric arcs and plasma physics.

Usually, high voltage laboratories involve tremendous cost. Hence, planning and layout have to be carefully done so that with the testing equipment chosen, the investment is not high and the maximum utility of the laboratory is made.

3. Classification of High Voltage Laboratories

High voltage laboratories, depending on the purpose for which they are intended and the resources available can be classified into three types.

(1) Small laboratories: A small laboratory is one that contains DC or power frequency test equipment of less than 10 kW/10 VA rating and impulse equipment of energy rating of about 10 kJ or less. Voltage ratings can be about 300 kV for AC, single unit or 500 to 600 kV AC for cascade units, ±200 to 400 kV DC and less than 100 kV impulse voltage.

(2) Medium size laboratories: Their main function will be for doing routine tests. Such a laboratory may initially contain a power frequency testing facility in the range of 200 to 600 kV depending on the ratings and the size of the equipment being manufactured and proposed to be- tested, such as cables, transformers etc. , but its kVA rating will be much higher (100 to 1000 kVA). The impulse voltage generator required would have a rating of 20 to 100 kJ or more.

(3) Large general—laboratories: This type of laboratories is meant to carryout testing

and undertakes research work and will contain almost all high voltage and high current test equipment and facilities. The building and equipment include the workshop, material handling equipment like cranes, ladders, etc. and large control and electric supply facilities (up to few kVA or MVA).

4. Layout of High Voltage Laboratories

The layout of a HV. laboratory is an important aspect for providing an efficient testing facility. Laboratory arrangements differ very much from a single equipment. Each laboratory has to be designed individually considering the type of equipment to be tested, the available space, other accessories needed for the tests, the storage space required, etc. Earthing, control gear, and the safety precautions require most careful consideration.

Laboratory Building

The building construction is not critical except where ionization tests are conducted. To minimize the floor loading problems and to simplify earthing arrangement, ground level location is preferred. The floor should withstand the loading imposed the equipment and test objects. Arrangements should be made to ensure that the laboratory is free from dust, draught, and excessive humidity. Laboratory window may require blackout arrangement for visual corona tests, etc. The control rod should be located in such a way as to include good overall view of the laboratory at test area. The main access door to the test area must accommodate the test equipment and the test object and have adequate interlocking arrangements and warning system to ensure safety to the personnel. A typical layout of a high voltage laboratory, accommodating a 1 MV AC. testing transformer and a 3 MV impulse generator, is shown in Fig. 5. 9. The dotted circles indicate the clearances necessary.

Words and Phases

acceptance test 验收试验，合格测验
lightning arrester 避雷装置，避雷针
accessory *n.* 附件，零件；
adj. 附属的，补充的
reactor *n.* 反应器，电抗器
test line 测试线路
plasma *n.* 等离子体，等离子区
cascade *n.* 串联，级联
control gear 控制装置（机构），操纵装置，自动调整仪
draught *n.* 气流
blackout *n.* 断电，灯火管制
control rod 控制棒，操纵杆
accommodate *v.* 供应，供给，向…提供，容纳
interlocking *adj.* 联锁的，相互连接的
Faraday cage 法拉第笼

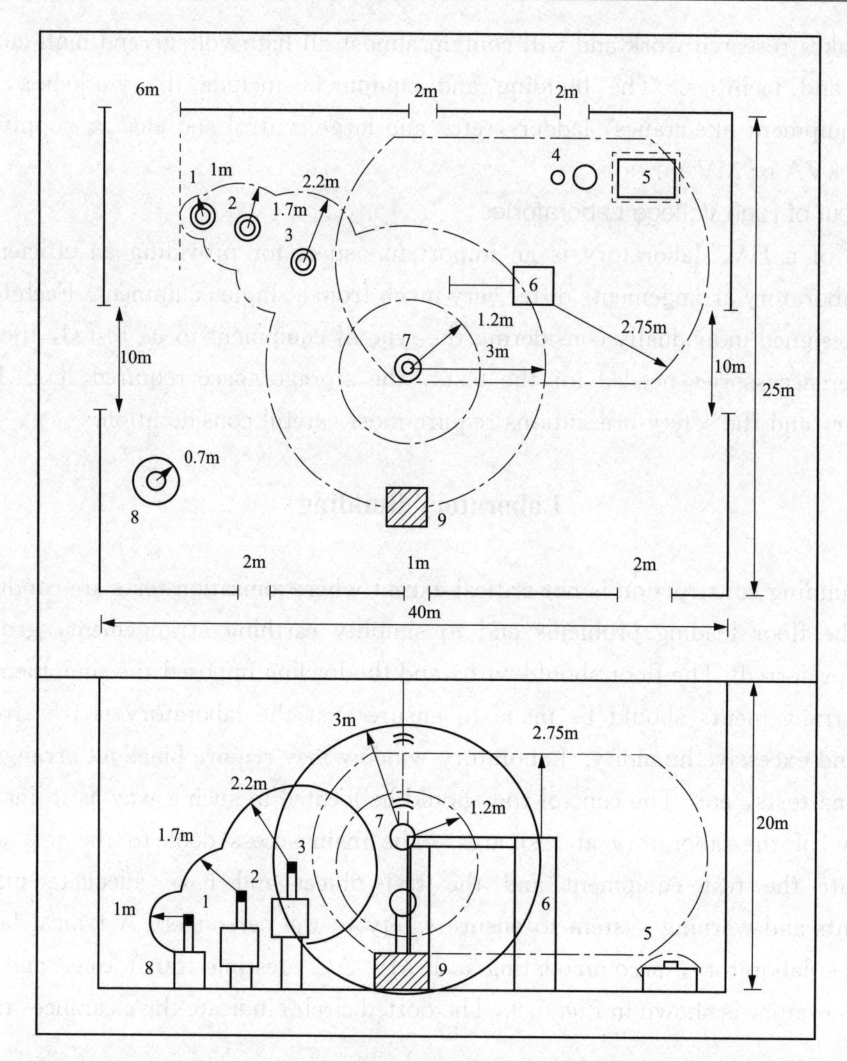

Fig. 5. 9 Layout of a typical high voltage laboratory with a 1 MV cascade transformer and 3 MV impulse generators 1～3—Cascade transformer set：1 MV；4 — DC charging unit：200kV；5—Impulse current generator：200 KA；6—Impulse voltage generator：3MV；7—Sphere gap：2 meters diameter；8—DC test set：300 kV；9—Control room Faraday cage：$5\times3.5\times3\ m^3$

Exercises

1. Answer the following questions according to the text

(1) What roles do the high voltage laboratories play in the extra high voltage transmission systems?

(2) What the test facilities are provided in high voltage laboratories?

(3) What research activities do high voltage laboratories include?

(4) What can high voltage laboratories be classified into?

(5) Where should the control rod be located in high voltage laboratories?

2. Translate the following sentences into Chinese according to the text

(1) Industrial and economic development in the present world demands the use of more and more electrical energy which has to be transported over long distances in large quantities.

(2) Sometimes, it is required to study problems associated with test lines and other equipment under natural atmospheric or pollution conditions, which cannot be done indoors.

(3) The building construction is not critical except where ionization tests are conducted.

(4) Arrangements should be made to ensure that the laboratory is free from dust, draught, and excessive humidity.

(5) The main access door to the test area must accommodate the test equipment and the test object and have adequate interlocking arrangements and warning system to ensure safety to the personnel.

3. Translate the following paragraph into Chinese

Transmission system planning criteria have been developed from the above planning principles and equipment ratings as well as from actual system operating data, probable operating modes, and equipment failure rates. These criteria are used to plan and build the transmission network with adequate margins to ensure a reliable supply of power to customers under reasonable equipment-outage contingencies. The transmission system should perform its basic functions under a wide range of operating conditions. Transmission planning criteria include equipment loading criteria, transmission voltage criteria, stability criteria and regional planning criteria.

附录A 科技英语阅读知识

1. 正常阅读速度监测方法

如何用最短的时间、最快的速度获取最丰富的信息，并对这些信息进行最有效的筛选，取得最佳效果，产生最好效应？对这些问题的圆满回答是：提高阅读速度、增强理解能力、改善记忆方式。

在进行系统的阅读技巧学习之前，首先应该对现阶段自身的阅读速度有所了解。具体检测方法是：所读文章单词数除以阅读所用的时间，算出每分钟能阅读多少单词的阅读速度（WPM）。

阅读速度计算公式

$$\text{每分钟单词数（WPM）}=\frac{\text{所读单词数}}{\text{时间}}$$

阅读能力参考表见附表A.1。

附表A.1 **阅读能力参考表**

读者水平	阅读速度（WPM）	理解程度
差	10～100	30%～50%
一般	200～240	50%～70%
专业人员	400～1000或1000以上	70%以上

2. 对快速阅读的误解

人们对快速阅读的普遍性误解有以下几点：①一次只能看一个单词。这种看法是错误的，是由于人们的凝视能力可以扩展，另外阅读不是理解单个的单词，而是整体意思。②阅读速度不可能超过每分钟500个单词。这种看法也是错误的，因为事实上每次凝视可以摄入6个单词，而且每秒钟可以凝视四次。这就意味着每分钟1000个单词的阅读速度是能达到的。③快速阅读者没法欣赏文章。这种观点还是错误的，因为快速阅读者能更多的理解所读的内容，能更专注的看材料，所以有更多的时间去回顾他认为特别有趣和重要的段落。④阅读越快注意力就越低。这种观点仍然是错误的，因为一般的阅读速度并不自然。造成这种错误观点的原因一是早期不完善的训练；二是缺乏眼睛和大脑能以各种可能的速度进行阅读等方面的知识。

3. 阅读的定义

阅读是个人对阅读时所牵涉的符号信息进行全部相互联系。它侧重于认知的视觉作用，阅读理解过程包含下述7个步骤：①辨识，读者对字母符号知识的掌握。②吸收，光从单词上反射并被眼睛吸收，再经由视神经传输到大脑的物理过程。③内部综合，是一种基本理解，特指将正被阅读信息的各个部分与其他相关部分连接起来的过程。④外部综合，该步骤包括分析、批评、鉴赏、选择与摒弃。是读者将其先前所学的整体知识用于正在阅读的新知识，并将二者适当的结合起来的过程。⑤保持，信息的基本储存过程。仅储存本身是不够的，必须用回忆加以完善。⑥记忆，指在需要时，从记忆库中提取所需要的信息，特别是指提取所需信息的能力。⑦交流，即信息的即时或最终使用过程。它包括非常重要的分支——思维。

4. 阅读方法

要想准确而又迅速地理解英语科技文章，必须有正确的阅读方法。阅读方法多种多样，每种阅读方法都有自己的优点和缺点，有自己的适用范围。我们应该根据自己的特点，针对阅读的具体要求和时间限制情况，选择合适的阅读方法。常用的阅读方法有：

（1）全读法。全读法是指一口气读完的阅读方法。这种阅读方法适用于阅读内容比较少、易懂、前后情节连贯的书籍和文章。

（2）逐段分读法。逐段分读法是指按照文章的段落逐段阅读的方法。这种阅读方法适用于阅读内容较深、篇幅较长的书籍和文章。

（3）精读法。精读法是指认真仔细而深入地阅读的方法。其关键在于读懂、读通、读活。必须做到深入掌握书的精髓，透彻地理解重点章节，深入思考并反复阅读书中的论点和结论，要发现问题和提出问题。精读法有利于吃透书，有利于举一反三。

（4）快读法。快读法是一种高速的阅读方法，其要点是从逐字逐词的认读，发展到逐行看读，再发展到数行及至十几行的扫读。要提高阅读速度就必须做到：①克服读书时发音或动嘴唇的毛病，养成直接看书的习惯。②养成整体认知的习惯。③掌握语言的运用规律。

（5）整体阅读法。整体阅读法是一种阅读前先浏览一遍的阅读方法，一般适用于阅读理论书籍。具体做法为：①整体把握。了解基本结构和主要部分，剖析文中各章节的结构和重点。其目的在于对全书有概要的了解，从而做到心中有数。②简读。一般以一事为一个单元，及时列出专题内容。

（6）三“W”读书法。三“W”读书法中的三“W”表示“What”“How”和“Why”三个英文单词。读书的过程是一个追求真理的过程，在这一过程中必然要经过从“什么”到“如何”再到“为什么”的全过程。这一过程也是读书知理所必须遵循的客观规律。这种阅读方法有助于搞清事物的来龙去脉及前因后果，有助于深入研究问题，有助于发展智力和创新。

5. 阅读技巧

要快速而准确地理解英语科技文章，既需要应用有效的阅读方法，又需要应用一定的阅读技巧。科技英语的阅读技巧很多，常用的阅读技巧包括：

（1）辨认重要事实。通过阅读，掌握英语科技文章中的重要事实和细节，猜测某一生词或短语的意义。辨认重要事实是阅读理解的最基本要求之一。文章中经常出现一些信号词来帮助我们迅速找到重要事实：例如 first，second，then，next，finally，also，in addition，besides，moreover，further，lastly 等单词或短语。

（2）得出合乎逻辑的结论。在阅读过程中，要能够批判地阅读文章，也就是要不断地提出问题：作者主要谈论什么内容？文中的事实或细节能否说明中心思想？所得到的结论是否建立在文中的事实之上？文章合乎逻辑吗？经过一系列的思维活动，读者才能得出正确的、合乎逻辑的结论。

（3）做出合理的判断。在阅读理解过程中，读者应对文中的事实、作者的观点和态度等做出合理的判断。为此，有三个步骤可供参考：①先理解文章中的基本事实或依据；②再评价这些事实或依据；③最后做出合理的判断。

(4) 进行严确的推理和概括。推理就是要根据已知的事实推断出未知的事实。在阅读过程中，要求理解文中没有直接说出的事实。为此，通常可采取如下三个步骤：①充分理解文中的某本事实；②分析这些事实；③进行正确的推理。

(5) 猜测词义。在阅读英语科技文章过程中，读者常常会遇到一些不认识、不熟悉的单词或短语。很多人习惯一遇到生词便查词典来确定这些词的意义，但这样做必然降低阅读速度。而且，有些疑难词在一般词典里也难以找到，或者虽然能找到，但词典中给出的解释在特定的语言环境下均不适用。另外，在一些特殊场合，例如闭卷考试时，根本不允许查词典。在这种情况下，我们只能运用自己已掌握的语言知识和一些语言技能来猜测词义。

(6) 注意对疑难句的分析。有时读者觉得一篇文章难懂往往不是因为内容或生词，而是句子结构。因此，能否简化难句对于阅读理解具有重要作用。在阅读过程中，读者可应用难句解析的方法来帮助理解。

难句解析是指运用分析的方法将结构复杂的句子化繁为简、化难为易。解析就是将整体分解为部分，将复杂的事物分解为简单的要素。运用此方法解剖难句，首先要了解结构复杂的句子的特点。在科技英语中，结构复杂的句子的一个重要特点是修饰语较多、较长。修饰语一般都是短语或从句，它们或是位于名词后面的定语从句、同位语从句或短语，或是位于动词后面的状语从句或短语。这类修饰语还可以一个套一个，甚至一连套上好几个，使句子结构变得复杂难懂。

附录B　科技英语翻译技巧

1. 翻译标准与过程

（1）翻译标准。翻译标准是翻译实践的准绳和衡量译文好坏的尺度。对于专业英语翻译，应达到“准确明白”“通顺严密”“简练全面”这三项标准。

1）准确明白：“准确”是说译文要准确无误地表达原文含义，不得有错；“明白”即要求译文不仅要忠实于原文而且应清楚明白地转达原文的意思，没有模糊不清之处。为此，必须正确理解原文，使译文不产生歧义。

2）通顺严密：“通顺”是指译文符合汉语的语法要求、修辞规则和通常习惯，使读者易看、易懂、易读；“严密”是说译文语言既要通畅又应严密，不要因为“通顺”而牺牲了原文的严密性。要达到此目的，译文的选词造句要符合汉语要求；译句中的词语之间、译文的译句之间要有呼应和关联，要逻辑清楚，层次分明，语气连贯，恰当地体现出原文的时态、语态和语气。

3）简练全面：“简练”是译文要简洁、精练，没有冗词费语，不重复啰嗦；“全面”是指译文不但要尽可能简练还应该力求全面，不能因简练而造成遗漏。这就要求翻译时在完全包容原文一切含义的前提下不受原文结构的限制，利用适当的技巧进行翻译，从而使译文简洁、精练、全面。

（2）翻译过程。科技英语翻译的过程主要为理解和表达的过程，大致可分为理解、表达、校核三个阶段：其中理解是前提，其主要内容是通读全文领略大意，明辨语法、弄清关系，纠合上下文推敲词意。表达是关键，一稿初译以忠实为主，二稿校核应注重逻辑，三稿定稿要润色文字。

1）理解阶段：其主要环节是辨明词义和语法关系，即正确判断英语句子中的语言现象，分析各个句子成分的逻辑关系和句子的语法结构，弄清单词、短语、从句的确切含义和句子所叙述的专业内容，并把前后句子联系起来理解，形成对原文的完整印象，真正掌握原文的内容和实质。

2）表达阶段：是在理解的基础上，以忠实于原意为前提，灵活地运用各种适当的翻译方法技巧，写出符合汉语规范、表达习惯以及翻译标准的译文。

3）校核阶段：是理解与表达的进一步深化，是对原文内容的进一步核实，对译文语言进一步推敲，进行必要的润色和修改，使译文符合标准规范。

为了达到翻译标准，需要利用各种翻译的方法和技巧。概括起来有次序的变更、词性的转换、词义的引申、用词的删减、句型的改造等。

2. 词义的确定

词是语言的最基本单位。在英语和汉语两种语言中，都存在着一词多用、一词多义的现象。翻译时，不可避免地要确定词义。词义的确定涉及词义的选择和引申两个方面。

（1）词义选择。

1）根据专业选择词义。在英汉两种语言中，同一个词在不同的学科领域或不同的专业往往具有不同的词义。因此，在选择词义时，应考虑阐述内容所涉及的概念是属于何种学科、何种专业。例如，cell一词含有“单元、细胞、蜂房、单人房间、电池”等意思。

请注意下列各个例句中cell的不同词义。

The nucleus is the information center of the *cell*.

细胞核是细胞的信息中枢。

He was imprisoned in a *cell*.

他被关在监狱的单人小牢房中。

In the center of the spacious workshop stood a *cell* with packets of block anodes.

在宽敞的车间中央有一个装有一组组阳极板的电解槽。

When the ends of a copper wire are joined to a device called an electric *cell*, a steady stream of electricity flows through the wire.

当把一根铜丝的两端连接到一种称为电池的电器上时，就会有稳定的电流流过该铜丝。

2）根据词类确定词义。英语中有些词在作为不同词类时有不同的意义，选择词义时，首先要判明一个词在原句中属哪个词类，再进一步确定其词义。现以like为例：

Like charges repel, unlike charges attract.

同性电荷相斥，异性电荷相吸。（形容词，作“相同的”解。）

In the sunbeam passing through the window there are fine grains of dust shining *like* gold.

在射入窗内的阳光里，细微的尘埃像金子一般在闪闪发亮。（介词，作“像”解。）

It is the atoms that make up iron , water, oxygen and the *like*.

正是原子构成了铁、水、氧等类物质。（名词，作“相同之物”解。）

He *likes* making experiments in chemistry.

他喜欢作化学实验。（动词，作“喜欢”解。）

Waves in water move *like* the waveform moves along a rope.

波在水中移动就像波形沿着绳子移动一样。（连词，作“像、如”解。）

3）根据上下文选择词义。根据上下文逻辑关系和语气连贯性来选择词义，这也是一种常用方法。例如：

It is impossible to predict detail the shape and *mechanism* of the robot slave. It might carry its computer and response *mechanism* around with it and also its source of power.

现在详细地预言这种机器人的形状和机理是不可能的。它可能自身带有电子计算机和反应装置及能源。（第一个mechanism译成“机理”，第二个mechanism译作“装置”）

Metallic iron contents were determined by electrochemical *solution* of the iron with copper nitrate *solution*.

金属铁的含量是通过用硝酸铜溶液对铁进行电化学溶解的方法加以测定的（第一个solution意为“溶解”，第二个solution则为“溶液”）

（2）词义引申。在英汉互译时，当原句中的一些单词或短语按词典的释义直接译出不符合译语语言规范和译文要求时，则可以根据上下文和逻辑关系，对原有词义作适当的引申。词义的引申主要有词义转译、词义具体化和词义抽象化三种手段。

1）词义转译。词义转译的目的是使译文语言流畅，文句通顺。例如：

The *beauty* of lasers is that they can do machining without ever physically touching the material.

激光的妙处就在于它能进行机械加工而不必实际接触所加工的材料。（不译成“美丽”）

The minute they started the test by shutting off the steam valves, their fate was *sealed*.

在他们关闭蒸汽阀以开始试验的这一刹那，他们的命运就决定了。（不译成“封闭”）

The difference between fusion and fission do*not stop there*.

聚变与裂变的差异不仅如此。（不译成“不停留在这里”）

Optics technology is one of the most sensational *developments* in recent years.

光学技术是近年来轰动一时的科学成就之一。（不译成“发展”）

2）词义具体化。词义的具体化是把原文中意义较笼统、抽象的词，根据汉语的表达习惯，引申为意义较明确、具体的词。

The major problem in fabrication is the control of contamination and *foreign materials*.

制造中的一个主要问题是如何控制沾染和杂质。（foreign materials 由“外界物质”具体化为“杂质”）

The foresight and *coverage* shown by the inventor of the process are most commendable.

这种方法的发明者所表现的远见卓识和渊博知识，给人以十分良好的印象。（coverage 由“范围”具体化为“知识范围”又转译为“渊博知识”）

3）词义抽象化。现代英语中，常常用一个表示具体形象的词来表示一种属性、一个事物或一个概念。英译汉时，一般要将其词义作抽象化的引申，或把词义较形象的词引申为较一般的词。

The *major contributors* in component technology have been in the semiconductor components.

元件技术中起主要作用的是半导体元件。[major contributors 由“主要贡献者”抽象化为“起主要作用的（因素）”]

There are three steps which must be taken before we *graduate from* the integrated circuit technology.

我们要完全掌握集成电路，还必须经过三个阶段。（graduate from 不译“毕业于”）

3. 词义增减翻译法

英汉两种语言，有不同的表达方式。在翻译过程中，要对语意进行必要的增减。在句子结构不完善，句子含义不明确或词汇概念不清晰时，需要对语意加以补足。反之，原文中的有些词如果在译文中不言而喻，就要省略一些不必要的词，使得译文更加严谨、明确。

（1）增补。增补是指翻译时出于明确表达原文含义和汉语语法修辞的需要，在译文中添加原文所没有的词语。按照增补的目的和作用，可将增补分为三类。

1）为明确原文词汇含义而增补。

a. 对抽象名词进行增补。这是针对不同的抽象名词的内在含义，在其后添加相应的名词（如作用、过程、现象、状态、能力、方法、工作、形式等），使其意义具体化。例如：

In rapid oxidation a flame is produced.

在快速氧化过程中会产生火焰。（对“氧化”增补“过程”）

Resonance is often observed in nature.

在自然界中常常观察到共振现象。（对“共振”增补“现象”）

b. 对普通名词进行增补。这是在普通名词后面添加相应的名词，以限定其概念所属的范畴。例如：

The arrows in the leads identify the materials.

引线的箭头标记着材料的类型。（对“材料”增补“类型”）

Light waves weaken as they spread out from a source.

光波从光源扩散时，强度减弱。（对“源”增补“光”；对“光波”增补“强度”）

c. 对形容词及名词定语进行增补。这是增补相应的词语，使译文语意明确。例如：

This receiver is indeed cheap and fine.

这台收音机真是物美价廉。（对“廉”增补“价”，对“美”增补“物”）

Since conduction is by both holes and electrons, the junction transistor is bipolar.

电子传导是靠空穴和电子两者进行的，因为结型晶体管是双极式的。（对“结”增补“型”；对“双极”增补“式”）

d. 对动词（包括非限定动词）进行增补。这也是增补相应的词语，使译文语意明确。例如：

The working instructions of this machine are to be gradually optimized during its practice.

这台机器的操作规程有待于在实际操作过程中逐步达到最佳化。（增补动词“达到”，将原文中的动词“optimize”转译成动宾结构“达到最佳化”）

Obviously, cool temperature slows down the action of bacteria.

显然，低的温度可使细菌的活动减慢。（增补“使”字，将原文中的动词“slows down”转译成汉语的兼语式“使活动减慢”）

We must make the phenomena clear.

我们必须把这些现象弄清楚。（增补“把”或“将”字，把原文中的动词转译成汉语的处置式）

A pulsed laser system has been tested on a satellite.

已经在人造卫星上对脉冲激光系统进行了试验。（增补“对”字，将宾语提前，并添加中性动词“进行”，将原文中的动词转译成动宾结构）

2）为表达原文语法概念而增补。

a. 增补表示复数概念的词。英语名词的复数形式一般不必译出。但在特定情况下，为明确表达原文的含义，须采用增补和重复的译法。例如：

Sensor switches are located near the end of each feed belt.

各传感开关位于每条给料皮带末端附近。（添加“各”字表示多个开关）

The three most important effects of an electric current are heating, magnetic and chemical effects.

电流的三种最重要的效应是热效应、磁效应和化学效应。（重复“效应”两次，表示有三种效应）

b. 增补表示动词时态的词。汉语里动词没有词形变化和相应的助动词，所以翻译时须添加相应的时间副词或助词，用来表示不同的时态。例如：

The electric motor converts electric energy into mechanical energy.

电动机可把电能转变为机械能。（对动词一般现在时有时可以增补“可”字）

That machine has run for three years.

那台机器已经运转了三年。(对动词完成式增补时间副词“已经”和助词“了”字)

c. 增补表示动词语气和语态的词。

The construction of the Taching oil field started in 1960.

大庆油田的建设是1960年开始的。(增补语气助词“的”字来表示陈述语气和强调的肯定句)

Is velocity a vector quantity?

速度是矢量吗?(增补语气助词“吗”字来表示疑问语气)

The force which pushes a bullet forwards is balanced by a reaction force.

推动子弹向前的力为反作用力所平衡。(增补“为…所”来表示被动语态)

Another factor to be considered is the speed.

另一个需要考虑的因素是速度。(增补“需要”来表示动词不定式被动语态作定语用时的潜在含义)

d. 增补连词。英语动词不定式及分词用作状语时,根据其内在含义,译成汉语时需要增添适当的连词(由于,因为,在…后,当…时,如果,只要,不论,尽管,为了,要(想),(从)而,以致等)。例如:

To make the bell stop ringing, you do not have to take both wires off the cell.

要使电铃停响,并不必把两根电线都跟电池断开。(增补连词“要”,表示目的状语)

Having been located, the short circuit can be put right.

短路在查出以后就能修好。(增补“在…以后”,表示时间状语)

e. 增补原文中省略了的词语。英语句子的某些成分如果已在前面出现过,当其再次出现时则往往予以省略。译成汉语时,一般需将其补出。例如:

High temperatures and pressures changed the organic materials into coal, petroleum and natural gas.

高温和高压把这些有机物变成了煤、石油和天然气。(原文 pressures 前面省略了 high,译文中予以补出、重复“高”字)

3)为满足汉语语法修辞要求而增补。

a. 增补主语。

It has been thought that radium radiations might be useful in curing various diseases.

人们认为镭的射线可以用来治疗各种疾病。(原文中 It 为形式主语,译文增补人称主语“人们”)

b. 增补宾语。

This is what we must revolve at first.

这就是我们必须首先解决的问题。(增补“问题”一词,它实质上是“解决”的宾语)

c. 增补谓语。

The sun gives us warmth and light.

太阳给我们带来了温暖和光明。(增补“带来了”三个字,补足谓语动词“gives”的语意)

d. 增补定语。

The plane twisted under me, trailing flame and smoke.

那架飞机在我下面螺旋下降,拖着浓烟烈火。(增补形容词“浓”利“烈”,勺“烟”和

"火"组成双音词)

e. 增补状语。

The speed is fast.

速度很快。(增补副词"很"字,以满足泽文语气需要)

The crowds melted away.

人群渐渐散开了。(增补副词"渐渐")

(2)省略。省略是根据汉语语法修辞习惯,将原文中的某些词语略去不译。如英语中的冠词、代词和连词(包括关系代词、关系副词等关联词),在中译文里件往可以省略。

1)省略冠词。

The atom is the smallest particle of an element.

原子是元素的最小的粒子。(the 和 an 省略)

A square has four angles. 正方形有四个角。(a 省略)

2)省略代词。

If you know the frequency, you can find the wave length.

如果知道频率,就能求出波长。(泛指的人称代词 you 省略不译)

Potential energy is the energy that a body has by virtue of its position.

位能是物体由于位置而具有的能量。(物主代词 its 省略不译)

3)省略动词。

When the pressure gets low, the boiling point becomes low.

气压低,沸点就低。(联系动词 get 和 become 省略)

This diode produces about nine times more radiant power than that one.

这只二极管(所产生)的辐射功率约为那只的九倍。(谓语动词 produce 省略)

4)省略介词。介词 for, in, by, of, with 等,在一定情况下可省略不译。

The first electronic computer was produced in our country in 1958.

我国第一台电子计算机是1958年生产的。(in 省略)

New boosters can increase the payload by 200%.

新型助推器能将有效载荷提高两倍。(by 省略)

5)省略连词。

An atom is so small that we cannot see it.

原子极小,我们看不见它。(that 省略)

High carbon steel is a kind of steel which tools are usually made of.

高碳钢是通常用来制造工具的一种钢。(which 省略)

6)省略不太必要的词。

The force required to turn a shaft depends on the length of the lever used.

使轴转动(所需)的力取决于(所用杆)臂(的)长(度)。(required 和 used 可予省略)

4. 词句转换翻译法

(1)词义的转换。翻译时,如有必要,可将原文中某个词的意义加以引申。这也是一种转换。例如:

The shortest distance between raw material and a finished part is precision casting.

把原料加工成成品的最简便方法是精密铸造。(将“最短距离”转换为“最简便方法”)

A network of highways was built from coast to coast.

建成了横跨大陆的公路网。(除将“从海岸到海岸”引申为“横跨大陆”外，全句倒译，原文中的状语转译为定语)

(2) 词性的转换。

Glass is more soluble than quartz.

玻璃的可溶性比石英大。(形容词转译为名词)

They laid special stress on raising the quality of electron devices.

他们特别强调提高电子器件的质量。(形容词转译为副词，名词转译为动词)

(3) 句子成分的转换。

There are three states of matter：solid，liquid and gas.

物质有三态：固态、液态和气态。(定语转译为主语；表语转译为宾语)

Gold has an advantage in that it cannot react with oxygen.

金子的优点是不与氧起反应。(主语转译为定语；宾语转译为主语)

(4) 句子结构的转换。

Open the key，and an induced current in the opposite direction will be obtained.

如果把电键断开，就会得到反向的感应电流。(并列复合句转译为条件句)

The simplest of electron tubes is diode which has two operating electrode-the heated cathode and the plate.

最简单的电子管是二极管，它有两个工作电极——热阴极和板极。(定语从句转译为并列分句)

(5) 表达方式的转换。

Sand may be carried many miles away by the wind.

风可以把沙土带到许多英里以外的地方。(被动语态转译为主动语态)

The articles in exhibition are not allowed to touch.

展品禁止触摸。(否定句转译为肯定句)

5. 否定句和被动句的译法

(1) 否定句的译法。

在英汉翻译，特别是专业英语翻译中，否定含义的理解和表达不容忽视。英语中表示否定的形式多种多样，如全部否定、部分否定、双重否定、意义否定（即形式肯定而内容否定）等。翻译中必须正确理解才能正确表达。

1) 全部否定。在英语中，全部否定是通过一些以字母“n”开头的否定词来表达的。这类否定词常用的有 no，none，not，never，nobody，nothing，nowhere，neither…nor 等。由这类词构成的全部否定与汉语的否定形式基本相同，一般可把表示否定的“不、无、非”等的词语动词连用。

There is no sound evidence to prove this.

没有可靠的证据可以证明这一点。

None of these metals have conductivity higher than copper.

这些金属中没有一种电导率比铜高。

The book was nowhere to be found.

那本书哪里也找不到了。

A gas has neither definite shape nor definite volume.

气体既没有一定的形状，也没有一定的体积。

2）部分否定。当英语的某些不定代词如all，both，every，each以及某些副词如often，always和否定词not连用时，就表示部分否定。

All these metals are not good conductors.

这些金属并不都是良导体。

Not all substances are good conductors of electricity.

并非所有的物质都是电的良导体。

Both of the instruments are not precise.

这两台仪器不都是精密的。

An engine may not always do work at its rated horse-power.

发动机并非总是以额定功率工作。

3）双重否定。在同一个句子中，两次运用否定手段就称为双重否定。由于双重否定实质上等于肯定，所以翻译时可译成双重否定句，也可反译成肯定句。

No area in the world is completely free of air pollution.

世界上完全没有被空气污染的地区是没有的。

As we a1l know，a1l 1ifecannot exist without water.

正如我们所知，没有水，任何生命都将不复存在。

There are no places left on the earth that the foot of man has not trodden.

地球上不存在绝无人迹的地方。

Nuclear radiation is not harmless to human being and other living things.

核辐射对人和其他生物都有害。

4）意义否定。有些英语句子，从形式或结构上看是肯定的，由于含有一些具有否定意义的词或词组，实际却是否定句，翻译时一般均应采用反译处理。

The motor *refused* to start.

马达开不动。

Another advantage of the *absence* of moving parts is that a transformer needs very *little* attention.

变压器没有运转部件的另一好处，在于它几乎不需要关注。

For a long time men thought that atoms were *indivisible*.

在很长一段时间内，人们都认为原子是不可分的。

There are some substances through which currents will *scarcely* flow at all.

有一些物质几乎完全不导电。

（2）被动句的译法。英语中被动语态使用范围很广，主语经常被省去，专业英语中更是如此。因为专业文献侧重叙事和推理，强调的是作者的观点和发明内容，而不是作者本人。汉语里很少使用被动语态，因此，在翻译时，应尽可能将英语的被动句译成汉语的主动句。视具体情况，也可保留被动句。

1）译成主动句。

It is clear that a body can be charged under certain condition.

很显然，在一定条件下物体能够带电。（原文中的主语 body 仍译作主语）

Electrical workshops may not be entered by unauthorized personnel.

未经许可的人员不得进入电气间。（把原文句中用作状语的介词短语的宾语 unauthorized personnel 译为主语，同时把原主语 electrical workshops 译为宾语）

If one of more electrons be removed, the atom is said to be positively charged.

如果原子失去一个或多个电子，我们就说该原子带正电荷。（原文中的主语 electrons 在译文中做宾语）

2）译成汉语被动句。译成汉语被动句时，除了用“被”外，还可使用“受”“由”“为…所”等。

Current will not flow continually, since the circuit is broken by the insulating material.

电流不能继续流动，因为电路被绝缘材料隔断了。

Resistivity is affected by temperature, moisture and structural defects.

电阻率受温度、湿度和结构缺陷的影响。

It is of logic circuit that all computers are made.

所有计算机都由逻辑电路组成。

Every charged object is surrounded by an electric field.

每个带电体都为电场所包围。

6. 从句的译法

英语的从句分为定语从句、主语从句、宾语从句、状语从句、表语从句以及同位语从句。由于英汉两种语言结构的不同，翻译时，应根据不同结构、不同含义采用不同译法。

（1）定语从句的翻译。定语从句一般由关系代词 that, which, who, as, but 等和关系副词 when, where, how, why 等引导，用来修饰主句中的某个名词（称为该关联词的先行词）。

1）合译法。把定语从句放在被修饰的词语之间，译成汉语的“‘的’字结构”，从而把定语从句和主句合译成汉语单句。

All substances which can conduct electricity are called conductors.

一切能导电的物质称为导体。

New electron tubes could be built that worked at much higher voltages.

如今制造的电子管，工作电压要高得多。（这一句定语从句是全句的重点，将从句顺序译成简单句中的谓语，从而突出从句的内容）

2）分译法。根据定语从句的不同情况，可将其翻译成并列分句、其他从句或词组等。

Mechanical energy is changed into electric energy, which in turn is changed into mechanical energy.

机械能转变为电能，而电能又转变为机械能。（译为并列分句）

Transformers cannot operate by direct current, which would burn out the wires in the transformer.

变压器不能使用直流电，因为直流电会烧坏其中的导线。（译为原因状语从句）

This, of course, includes the movement of electrons, which are negatively charged particles.

当然，这包括电子（带负电的粒子）的运动。（定语从句在本句的翻译中转变为词组）

(2) 主语从句的翻译。

1)"的"字结构。以 what，that，who，where，whatever 等代词引导的主语从句，可以将从句翻译成"的"字结构。

What a motor does is to change electrical energy into mechanical driving power.

电动机所起的作用就是把电能转变成机械能。

Where the power station will be built is under discussion.

发电厂建在什么地方正在讨论之中。

2) 顺译法。如果主语从句是在 It＋谓语＋主语从句的结构中，翻译时也可以采取顺译法，先译主句（译成无人称句），后译主语从句。

It is reported the old power network will be transformed.

据报道，旧的电网将被改造。

It is of course required that this internal currents loss be at a minimum.

当然要求这种内部电流的损失达到最小。

(3) 宾语从句的翻译。由 that，what，how，where 等词引导的宾语从句一般按照原文顺序译出，即顺译法。

Note that increasing the length of the wire increase its resistance.

应注意，增加导线长度就会增加导线的电阻。

I don't know if we can get the transformer repaired on time.

我不知道我们能否按时修好变压器。

(4) 状语从句的翻译。

1) 顺译法。英语大部分状语从句的位置可以放在主句之前，也可以在主句之后，还可以在主句的主谓语之间。当这些从句位于主句之前时，可以按原顺序译出。

Where you are on the earth，you are not free from the force of gravity.

不管你在地球上的什么地方，都摆脱不了重力。（地点状语从句）

In order that an electric current may flow in a circuit the latter must be complete and must consist of materials which are electrical conductors.

要使电流流过线路，线路就必须完整且由导体组成。（目的状语从句）

2) 倒译法。当状语从句位于主句之后时，一般应逆着原文顺序译出。

Both plants and animals could not exist where there were no sunlight，water and air.

在没有阳光、水和空气的地方，动、植物是不能生存的。（地点状语从句）

The death-rate could begin to rise again because medicine cannot keep up with the continued rise in population.

由于医疗水平赶不上人口的持续增长，因而死亡率可能再度上升。（原因状语从句）

(5) 表语从句的翻译。翻译表语从句一般都可照原文顺译，有时可译为"……是……"的句式。

What you should know is that the condensers will rest on independent supports below the turbine.

你应该知道的是冷凝器要安装在汽轮机下单独的支座上。

The question is what step we should take next.

问题在于下一步我们该采取什么措施。

That is how voltage is measured.

那就是测量电压的方法。

（6）同位语从句的译法。先翻译从句，即从句前置。

This is a universally accepted principle of international law that the territory sovereignty does not admit of infringement.

一个国家的领土不容侵犯，这是国际法中尽人皆知的准则。

Despite the fact that comets are probably the most numerous astronomical bodies in the solar system aside from small meteor fragments and the asteroids，they are largely a mystery.

在太阳系中除小片流星和小行星外，彗星大概是数量最多的天体了，尽管如此，它们仍旧基本上是神秘莫测的。

7. 长句的译法

翻译长句子时，必须分析清楚原文的结构关系，在理解原文的基础上，分清主次，按照时间概念和逻辑顺序重新按汉语习惯加以组合处理，译出的句子既要准确的表达原文，又要符合汉语的习惯。常用的译法有顺译法、倒译法和分译法。

（1）顺译法。顺译法是指基本上保留原文的语法结构，依照原文顺序译出。一般地，当原文的语法结构和时间顺序与汉语相同时，采用顺译法。

In a cross-field machine，however，the temporary increase in speed would result in a great quadrature current for the same control field excitation，and owing to hysteresis the output voltage when the speed was restored to normal would not be the same as would have been obtained if the speed had remained constant throughout.

但是，在交磁发电机中，转速的暂时增加在相同的控制磁场绕组励磁电流下会引起较大的交轴电流，而由于磁滞效应使转速恢复到额定值时的输出电压，将不同于转速始终保持常数时所得到的电压值。（主体结构是并列句，后一个分句含一个由 when 引导的定语从句和一个由 as 引出的定语从句，以及一个由 if 引出的条件句）

（2）倒译法。有些英语长句子的表达顺序可能与汉语的表达顺序正好相反，这种长句子在翻译时就得用倒译法。

You must keep in mind the symbols and formulae，definitions and laws of physics，no matter how complex they may be，when you come in contact with them，in order that you may understand the subject better and lay a solid foundation for further study.

为了更好地了解物理学，并为进一步学习打下坚实的基础，当你接触到物理学的符号、公式、定义和定律的时候，不论它们是多么复杂，你也必须记住。

（3）分译法。在英语长句子中，有时主句与从句，或短语之间关系并非十分紧密，且有时翻译时为使译文更符合汉语的表达习惯，使行文方便，往往可以将繁杂的长句子分开翻译。分译时，为了使译文语意连贯，常常可以适当地增译或者省译某些词语。

The diode consists of a tungsten filament，which gives off electrons when it is heated，and a plate toward which the electrons migrate when the field is in the right direction.

二极管由一根钨丝和一个极板组成，钨丝受热时放出电子；当电场方向为正时，这些电子就移向极板。

附录C 英语写作方法

1. 应用文写作

(1) 简历 (Resume)。简历是个人经历的文字记载。撰写个人简历是为了一定的目的，向特定的读者介绍个人的有关情况。

简历写作要求做到实事求是，简明扼要，语言正确，格式规范。简历的内容因人、因事而各有不同。但为了比较全面地反映个人的基本情况，一个较为完整的简历大体包括：个人情况、教育状况、工作经历、出版物、推荐人等项目。作者可根据实际需要选择增补。

1) 个人情况 (Personal)。个人情况通常包括个人的姓名、性别、出生日期、出生地点、国籍、婚姻状况、配偶资料、健康状况等。有时，还需注明常住地址、通信地址、电话及电子邮箱等。根据特定要求，有时甚至还需写明个人的身高、体重以及爱好等。

2) 教育状况 (Education)。教育状况主要写明何时何地何学校毕业、何时何处被授予何种学位。根据需要，还应当写明自己的外语程度。一般最高学历或最近学历先写，时间安排成倒叙的顺序。

3) 工作经历 (Experience)。工作经历的栏目除 Experience 外，还可视具体情况写成 Working，Experience，Research Experience，Teaching Experience，Professional Experience，Employment 等。一般地说，工作经历应包括工作单位、工作时间、工作内容、工作成效以及担任的职务等。写作时，依据简历的用途和对方的具体要求确定详略的程度。叙述的顺序也可按由近及远的倒序方法。

4) 出版物 (Publications)。出版物是个人专业能力和学术水平的一个重要体现。若以专著出版，以 Books 为题列出书籍的名称、出版者、出版年份、总页数等。论文另立一项 Papers，列出主要论文的作者（包括合作者）、标题、刊名、刊期、页码等。翻译的文献以 Translation 为题列出。出版物的书写体例一般按写作参考文献的规定。出版物一般按时间顺序由远及近写为好。

(2) 产品说明书。产品说明书是生产单位向用户介绍产品的特点、性能、用途、使用、保养维修等的文字说明。目的是帮助广大用户了解产品，指导用户正确使用产品。产品说明书要求语言平实、简洁、准确、条理分明，注重内容的科学性和实用性。

产品说明书的结构一般由标题、前言、正文、落款等部分组成。

1) 标题一般采用产品名称，或产品名称加文种（“说明书”或“使用说明书”）。

2) 前言也称概述。用简练的语言概括介绍产品的主要特点，适用范围等。

3) 正文是产品说明书的主体部分，一般应包括以下几方面的内容：

a. 产品的研究、开发及制作工艺。

b. 产品的性质、性能、特征、特点等。

c. 产品的适用范围和用途等。

d. 产品的安装及使用方法。

e. 产品的保养、维修方法及注意事项等。

f. 产品的工作原理、主要技术指标等。

根据不同的产品，产品说明书的内容也各有侧重。

4）落款，在正文后面详细写出生产厂家的通信地址及邮政编码、电话号码等。

产品说明书正文的写作形式多种多样，有条文式、表格式等，使用的说明方法也有很多，可以采用定义说明、图表说明、分类说明、数字说明等。

对于内容较为复杂的产品说明书，常印成小册子或手册，由封面、目录、前言、正文、封底及插图等组成。

2. 科技论文的写作要求和分类

(1) 写作要求。科技论文写作是一件十分严肃的事情。一篇科技论文既体现了作者的思维，也体现了作者的学风。科技论文的类型多种多样，不同类型科技论文的写作要求既有共同之处，又有不同之处。对科技论文写作的共同要求一般是思维清晰、表达准确、论证严密、内容客观、叙述完整、语言简练。

(2) 分类。科技论文是记录科技进步的历史性文件，也是当今科技交流的主要信息源。科技论文主要包括学位论文、学术论文和科技报告三大类。

1）学术论文。学术论文是论述创造性研究成果的书面文件，是创造某种原理或将某种已知原理应用于实际取得新进展的科学总结，或是某一学术课题在实验性、理论性、观测性上具有新的科研成果或创新见解和知识的科学记录。学术论文通常用于学术会议上宣读、交流或讨论，或在学术期刊上发表。

2）学位论文。学位论文是作者提出申请授予学位时向学位授予机构提交的供评审用的科技论文。学位一般分为学士、硕士和博士三级。

学士是最低层次的学位，因此对学士学位论文的要求也最低。通常只要求学士学位论文能体现作者具有从事科学研究工作或担负专门技术工作的初步能力。

3）科技报告。科技报告是描述一项科研成果和进展，或一项技术装备研制试验和评价结果，或论述某一科学技术问题的现状和发展趋势的文件。科技报告通常具有一定的保密期限和保密级别，保密级别通常分为一般、内部、秘密和绝密四个等级。

3. 科技论文的写作

科技论文一般由以下这几部分组成：论文题目（Title）；作者姓名（Author's name or Authors' names）；摘要（Abstract）；正文（Main body）；参考文献（Bibliography）。

(1) 题目。论文题目是论文篇首的文字，一般应写在文章首行的正中央。题目不宜过长（一般不超过两行）。如果题目过长，一行之内排不开，则可排成倒金字塔状，但是，词或短语不要分开写。题目中，除介词、连词外，每个词的词首需大写。

题目是论文内容的高度概括，应充分反映论文的基本内容和特色。其基本要求是确切、简练、醒目。一般采用各种名词性结构，即中心词（名词）+修饰语（介词短语、分词短语、不定式短语等）。如：

An Digital Processor of Neural Networks.

一种基于神经网络的数字信息处理器。

To Apply (Applying) the Principle of Phrase Comparison of Fault Component to Transformer Protection.

应用故障分量比相原理保护变压器。

论文标题中有些冠词可以省略。

Summary on Development of Switched Reluctance Machine.

开关磁阻电机发展综述（在 Summary 前省略了定冠词 The）。

论文题目中常常出现的词汇归纳如下：summary（综述），development（发展），application（应用），comparison（比较），study（设计），research（研究），experimental research（实验研究），application research（应用研究），design（设计），analysis（分析），method（方法），principle（原则），investigation（调查），discussion（讨论）。

（2）作者姓名。作者姓名及所在单位编排在题目下面。作者多于一个时，应按字母顺序排，如果论文主要是某一作者完成的，则将该作者排在第一位。科技论文的作者姓名的英文应写全名，包括名（given name）、姓（surname）和中间缩写名（middle initial）。此外还要附上作者的专业技术职称、作者的最高学位工作单位与邮政编码，以便读者与作者联系。如：

Li Junqing　　Li Heming

North China Electric Power University，Baoding，China

（3）摘要。摘要是从文章中摘录的要点，是论文内容的简洁和精确的叙述。摘要内容应包括研究的目的、方法、结果和结论。大多数英文摘要的字数应掌握在 250 个单词以内。短文的摘要更应少于 100 个单词。摘要的第一个句通常是一个带主题性的句子，用以归纳未被论文题目所表达的必要内容。摘要中时态形式出现最多的是一般现在时，其次是现在完成时。

在科技论文的英文摘要之后要列出本文的关键词。其目的是便于选读和计算机检索存储的需要。关键词必须精选能够代表主要内容的术语。关键词词数规定为 3～5 个。关键词通常列于简要之后，另起一行开始。关键词或关键词组之间应有 2～3 个空格或者分号隔开。

摘要中常用到的词汇归纳如下：

be described（描述），be introduced（介绍），be expounded（详加解释），be discussed（讨论），be analyzed（分析），be offered（提供），be pointed out（指出），be inferred（推导），be proposed（提出），be developed（开发），be tested（验证），be used（使用），be built（建立），be adopted（采用），be applied（应用），be investigated（调查），be proved（证明），be compared（比较），be concluded（得出结论），be improved（改进），be presented（提出），be studied（研究），be shown（表明），be given（给出），be examined（检查），be reviewed（回顾），be outlined（概括）。

下面是一篇论文的摘要。

Abstract

Development，application and structure of the switched reluctancesystem are briefly described. Configuration of stator and rotor，excitation mode and working principle for various switched reluctance machine are expounded. Design methods about the switched reluctance machine are discussed and its advantage，disadvantage and application scope are introduced in detail. Finally，its existing problems and developing trends are analyzed.

摘要

简述了开关磁阻电机的发展及其应用领域，介绍了开关磁阻电机系统的构成。系统阐述了开关磁阻电机各种结构形式中，定、转子的组成，励磁方式及其工作原理。论述了开关磁阻电机的设计方法，对每种设计方法的优缺点及其适用范围做了较详细的介绍。最后分析了开关磁阻电机设计理论、分析方法中存在的问题及其发展趋势。

（4）正文。正文是科技论文的核心部分，内容应充实，要阐明作者研究成果或所提出理论、方法的基本内容、理论依据、正确性、实现方法、仿真或实验的验证方法及其结果分析。要求推理论证严密，论述简明准确、分析与介绍条理清晰而重点突出，全文层次清楚、结构严谨、合乎逻辑。

科技论文在体裁上属说明文。说明文的目的是解释事物，说明道理。解释事物时可以通过定义、分类、举例、比较和对比，分析其因果关系和提供数据等手段。因此，在构思段落时，可以有效地运用这些方式。

1）定义。定义是揭示概念的内涵或者词语意义的方法。在解释事物时，往往需要给出一些概念或者词语明确的定义。

A substation is an assemblage of equipment for the purpose of switching of changing or regulating the voltage of electricity.

变电站是把一些设备组装起来，用以切断或接通，改变或者调整电压。[这一段是给变电站（substation）下了个定义]

在下定义时，常要用到以下这些词和短语：be，be defined as，be known as，be called，mean，refer to，involve，be concerned with。

2）分类。分类是把具有共同特性的有关事物分成属类。它的目的是系统化，使有共同特征的事物系统的组合在一起。在科技论文写作中，分类是非常有效的一种组织文章的策略。

Main classifications of the substations are as follow：Step-up substation，primary grid substation，secondary substation and distribution. 变电站主要分为以下几类：升压变电站，主网变电站，二次变电站和变电站。（这一段对变电站进行了分类）

在对事物进行分类时常要用到这些词语：class，group，category，kind，sort，be classified into，be divided into，group into，be made up of。

3）举例。举例是使用例子说明要点。它是最常见，也是最有效的一种说明方式。举例时，常用的词语有：like，an example of，such as，be illustrated by，for example，as illustrated by，for instance。

4）比较与对比。比较是确定事物同异关系的思维过程和方法，即根据一定的标准把被此有某种联系的事物加以对照，从而确定其相同与相异之处，以便对事物进步初步的分类。对比则是把两种不同的事物作对照，互相比较。

常用的词语有：comparison，contrast，both，similar，difference，different，like，same，on the other hand，on the contrary，however，instead。

5）因果分析。因果分析用于阐述和分析某事件产生的原因和结果。常用到的词语有：as，because，since，according，consequently，hence，therefore，thus，so，as a result，account for。

（5）结论和参考文献。

1）结论或总结。结论是对论文的最终的、总结式的评价和提高，是让读者对本文的精髓、意义、新思想、新方法以及验证结果有一个全貌的了解。应做到准确、完整、明确、精练。在结论中还要简单介绍研究成果可以推广的应用领域，对进一步研究提出建议、研究设想、尚需解决的问题等。

（2）参考文献。在科技论文的写作中，参考文献占有重要地位。参考文献不仅是为方便读者查阅，它本身也反映了作者的治学作风是否严谨，对他人的学术成果是否尊重。因此，科技工作者在撰写学术论著时，一定要认真地处理好参考文献。参考文献的顺序一般需与其在论文中被引用的顺序一致。

根据引用参考文献的类型分别采用以下顺序编排：

1）著作 作者. 译者. 书名. 出版地：出版社，出版年：起-止页.

2）期刊论文 作者. 题（篇）名. 出版年，卷号（期号）：起-止页.

3）会议论文 作者. 题（篇）名. 文集名. 会议名，会址，开会年：起-止页.

4）学位论文 作者. 题（篇）名. ［博（硕）士学位论文］. 授学位学校. 授学位年.

5）网上搜索 网址. 论文作者. 论文名称.

4. 电气工程及其自动化专业英文主要期刊一览表

IEEE Transactions on Power Delivery IEEE 输电汇刊

IEEE Transactions on Power Systems IEEE 电力系统汇刊

IEEE Transactions on Industry Applications IEEE 工业应用汇刊

IEEE Power Engineering Review IEEE 动力工程评论

IEEE Transactions on Energy Conversion IEEE 能源转换汇刊

IEE Proceeding-Generation Transmission and Distribution IEEC 辑：发电、输电与配电

IEEE Transactions on Circuits and Systems IEEE 电路与系统汇刊

IEE Proceedings G-Circuits，Devices and Systems IEEG 辑：电路、器件和系统

IEEE Transactions on Dielectrics and Electrical Insulation IEEE 介质与电气绝缘汇刊

EPRI Journal 美国电力研究会汇刊

Electrical World 电世界

Power Engineering 动力工程

Energy Policy 能源政策

International Water Power and Dam Construction 水力发电与坝工建设

Asian Electricity 亚洲电力

Transmission and Distribution World 输电与配电世界

Electric Power System Research 电力系统研究

Electric Light and Power 电气照明与动力

参 考 文 献

[1] [美] 菲茨杰拉德，[美] 小金斯利，[美] 尤曼斯. Electric machinery. 6 版（影印本）. 北京：清华大学出版社，2003.

[2] 祝晓东，张强华，古绪满. 电气工程专业英语实用教程. 北京：清华大学出版社，2006.

[3] http://www.etrema-usa.com.

[4] Designing for Board Level Electromagnetic Compatibility，T. C. Lun，Freescale Semiconductor Application Note，国际电器产品与可靠性年会论文集，2003.

[5] Theodore Wildi. Electrical Machines，Drives，and Power Systems. Fifth Edition. 北京：科学出版社，2002.

[6] Ned Mohan，Tore M. Undeland，William P. Robbins. Power Electronics—Converters，Applications，and Design. Third Edition. 北京：高等教育出版社，2004.

[7] Bimal K. Bose. Modern Power Electronics and AC Drives. 北京：机械工业出版社，2003.

[8] J. Duncan Glover，Mulukutla. S. Sarma. Power system Analysis and Design. 北京：机械工业出版社，2004.

[9] Bergen. A. R. Power Systems Analysis. Second Edition. 北京：机械工业出版社，2004.

[10] 刘健. 电力英语阅读与翻译. 2 版. 北京：中国水利水电出版社，2003.

[11] 苏小林，顾雪平. 电气工程及其自动化专业英语. 北京：中国电力出版社，2005.

[12] M S Naidu，V Kamaraju. High Voltage Engineering. 2 Edition. Tata McGraw-Hill Publishing Company，1995.

[13] Mo-Shing Chen，K. C. Lai，and Rao S. Thallam. High-Voltage Direct-Current Transmission. 2000 by CRC Press LLC.

[14] 宇德明. 科技英语阅读与写作. 北京：中国铁道出版社，2002.

[15] 阎庆甲，阎文培. 科技英语翻译方法（修订版）. 北京：冶金工业出版社，1992.

[16] 阎庆甲. 科技英语翻译手册. 郑州：河南科学技术出版社，1986.

[17] 王运. 实用科技英语翻译技巧. 北京：科学文献技术出版社，1992.

[18] 王泉水. 科技英语翻译技巧. 天津：天津科学技术出版社，1991.

[19] 韩其顺，王学铭. 英汉科技翻译教程. 上海：上海外语教育出版社，1992.

[20] 郑仰成. 电力科技英语翻译方法与技巧. 北京：中国水利水电出版社，2002.

[21] 郑仰成. 电力英语应用文写作. 北京：中国水利水电出版社，2003.

[22] 袁道之，素金梅. 科技英语写作指南. 北京：电子工业出版社，1988.

[23] 王建武，李民权. 科技英语写作——写作技巧·范文. 西安：西北工业大学出版社，2000.

[24] J. F. Gieras. Permanent Magnet Motor Technology：Design and Applications，3rd edition，2010 by Taylor and Francis Group，LLC.

[25] T. A. Lipo. Introduce to AC machine design. 2017 by The Institute of Electrical and Electronics Engineers，Inc.

[26] Callister William. Materials science and engineering：An introduction. Seven edition. New York：John Wiley & Sons，Inc.，2007.

[27] B. D. Cullity. Fundamentals of Silicon Carbide Technology. John Wiley & Sons，2014.

[28] F. Wang and Z. Zhang. Overview of Silicon Carbide Technology：Device，Converter，System，and Application. CPSS Transactions on Power Electronics and Applications，2016，1（1）：13-32.

[29] M. Trzynadlowski. Introduction to Modern Power Electronics. John Wiley & Sons, 2016.

[30] X. Zhou, P. L. Wong, P. Xu, F. C. Lee, and A. Q. Huang. Investigation of Candidate VRM Topologies for Future Microprocessors. IEEE Trans. Power Electron., vol. 15, no. 6, pp. 1172-1182, Nov. 2000.

[31] Liang T J, Tseng K C. Analysis of integrated boost-flyback step-up converter. IEE Proceedings - Electric Power Applications, vol. 152, no. 2, pp. 217-225, Mar. 2005.

[32] Y. Gu, W. Li, Y. Zhao, B. Yang, C. Li, and X. He, Transformerless inverter with virtual DC bus concept for cost-effective grid-connected PV power systems. IEEE Trans. Power Electron., vol. 28, no. 2, pp. 793-805, Feb. 2013.

[33] D. Meneses, F. Blaabjerg, O. Garcia, and J. A. Cobos. Review and comparison of step-up transformerless topologies for photovoltaic AC-module application. IEEE Trans. Power Electron., vol. 28, no. 6, pp. 2649-2663, Jun. 2013.

[34] S. B. Kjaer, J. K. Pedersen, and F. Blaabjerg. A review of single-phase grid-connected inverters for photovoltaic modules. IEEE Trans. Ind. Appl., vol. 41, no. 5, pp. 1292-1306, Sept-Oct. 2005.

[35] N. Hatziargyriou, H. Asano, R. Iravani, et al. Microgrids. IEEE Power and Energy Magazine, 2007, 5 (4): 78-94.

[36] G. W Arnold. Challenges and Opportunities in Smart Grid: A Position Article. Proceedings of the IEEE, 2011, 99 (6): 922-927.